DIGITAL INSTRUMENTATION AND CONTROL SYSTEMS AND OTHER ADVANCED DIGITAL TECHNOLOGIES FOR ENHANCING NUCLEAR POWER PLANT PERFORMANCE

The following States are Members of the International Atomic Energy Agency:

AFGHANISTAN
ALBANIA
ALGERIA
ANGOLA
ANTIGUA AND BARBUDA
ARGENTINA
ARMENIA
AUSTRALIA
AUSTRIA
AZERBAIJAN
BAHAMAS, THE
BAHRAIN
BANGLADESH
BARBADOS
BELARUS
BELGIUM
BELIZE
BENIN
BOLIVIA, PLURINATIONAL
  STATE OF
BOSNIA AND HERZEGOVINA
BOTSWANA
BRAZIL
BRUNEI DARUSSALAM
BULGARIA
BURKINA FASO
BURUNDI
CABO VERDE
CAMBODIA
CAMEROON
CANADA
CENTRAL AFRICAN
  REPUBLIC
CHAD
CHILE
CHINA
COLOMBIA
COMOROS
CONGO
COOK ISLANDS
COSTA RICA
CÔTE D'IVOIRE
CROATIA
CUBA
CYPRUS
CZECH REPUBLIC
DEMOCRATIC REPUBLIC
  OF THE CONGO
DENMARK
DJIBOUTI
DOMINICA
DOMINICAN REPUBLIC
ECUADOR
EGYPT
EL SALVADOR
ERITREA
ESTONIA
ESWATINI
ETHIOPIA
FIJI
FINLAND
FRANCE
GABON
GAMBIA, THE

GEORGIA
GERMANY
GHANA
GREECE
GRENADA
GUATEMALA
GUINEA
GUYANA
HAITI
HOLY SEE
HONDURAS
HUNGARY
ICELAND
INDIA
INDONESIA
IRAN, ISLAMIC REPUBLIC OF
IRAQ
IRELAND
ISRAEL
ITALY
JAMAICA
JAPAN
JORDAN
KAZAKHSTAN
KENYA
KOREA, REPUBLIC OF
KUWAIT
KYRGYZSTAN
LAO PEOPLE'S DEMOCRATIC
  REPUBLIC
LATVIA
LEBANON
LESOTHO
LIBERIA
LIBYA
LIECHTENSTEIN
LITHUANIA
LUXEMBOURG
MADAGASCAR
MALAWI
MALAYSIA
MALDIVES
MALI
MALTA
MARSHALL ISLANDS
MAURITANIA
MAURITIUS
MEXICO
MONACO
MONGOLIA
MONTENEGRO
MOROCCO
MOZAMBIQUE
MYANMAR
NAMIBIA
NEPAL
NETHERLANDS,
  KINGDOM OF THE
NEW ZEALAND
NICARAGUA
NIGER
NIGERIA
NORTH MACEDONIA
NORWAY

OMAN
PAKISTAN
PALAU
PANAMA
PAPUA NEW GUINEA
PARAGUAY
PERU
PHILIPPINES
POLAND
PORTUGAL
QATAR
REPUBLIC OF MOLDOVA
ROMANIA
RUSSIAN FEDERATION
RWANDA
SAINT KITTS AND NEVIS
SAINT LUCIA
SAINT VINCENT AND
  THE GRENADINES
SAMOA
SAN MARINO
SAUDI ARABIA
SENEGAL
SERBIA
SEYCHELLES
SIERRA LEONE
SINGAPORE
SLOVAKIA
SLOVENIA
SOMALIA
SOUTH AFRICA
SPAIN
SRI LANKA
SUDAN
SWEDEN
SWITZERLAND
SYRIAN ARAB REPUBLIC
TAJIKISTAN
THAILAND
TOGO
TONGA
TRINIDAD AND TOBAGO
TUNISIA
TÜRKİYE
TURKMENISTAN
UGANDA
UKRAINE
UNITED ARAB EMIRATES
UNITED KINGDOM OF
  GREAT BRITAIN AND
  NORTHERN IRELAND
UNITED REPUBLIC OF TANZANIA
UNITED STATES OF AMERICA
URUGUAY
UZBEKISTAN
VANUATU
VENEZUELA, BOLIVARIAN
  REPUBLIC OF
VIET NAM
YEMEN
ZAMBIA
ZIMBABWE

The Agency's Statute was approved on 23 October 1956 by the Conference on the Statute of the IAEA held at United Nations Headquarters, New York; it entered into force on 29 July 1957. The Headquarters of the Agency are situated in Vienna. Its principal objective is "to accelerate and enlarge the contribution of atomic energy to peace, health and prosperity throughout the world".

IAEA NUCLEAR ENERGY SERIES No. NR-T-1.22

DIGITAL INSTRUMENTATION AND CONTROL SYSTEMS AND OTHER ADVANCED DIGITAL TECHNOLOGIES FOR ENHANCING NUCLEAR POWER PLANT PERFORMANCE

INTERNATIONAL ATOMIC ENERGY AGENCY

VIENNA, 2026

IAEA Library Cataloguing in Publication Data

Names: International Atomic Energy Agency.
Title: Digital instrumentation and control systems and other advanced digital technologies for enhancing nuclear power plant performance / International Atomic Energy Agency.
Description: Vienna : International Atomic Energy Agency, 2026. | Series: IAEA nuclear energy series, ISSN 1995-7807 ; no. NR-T-1.22 | Includes bibliographical references.
Identifiers: IAEAL 25-01768 | ISBN 978-92-0-114425-6 (paperback : alk. paper) | ISBN 978-92-0-114225-2 (pdf) | ISBN 978-92-0-114325-9 (epub)
Subjects: LCSH: Nuclear power plants — Technological innovations. | Nuclear power plants — Design and construction. | Nuclear power plants — Management. | Nuclear power plants — Software.
Classification: UDC 621.039.56 | STI/PUB/2116

FOREWORD

The IAEA's statutory role is to "seek to accelerate and enlarge the contribution of atomic energy to peace, health and prosperity throughout the world". Among other functions, the IAEA is authorized to "foster the exchange of scientific and technical information on peaceful uses of atomic energy". One way this is achieved is through a range of technical publications including the IAEA Nuclear Energy Series.

The IAEA Nuclear Energy Series comprises publications designed to further the use of nuclear technologies in support of sustainable development, to advance nuclear science and technology, catalyse innovation and build capacity to support the existing and expanded use of nuclear power and nuclear science applications. The publications include information covering all policy, technological and management aspects of the definition and implementation of activities involving the peaceful use of nuclear technology. While the guidance provided in IAEA Nuclear Energy Series publications does not constitute Member States' consensus, it has undergone internal peer review and been made available to Member States for comment prior to publication.

The IAEA safety standards establish fundamental principles, requirements and recommendations to ensure nuclear safety and serve as a global reference for protecting people and the environment from harmful effects of ionizing radiation.

When IAEA Nuclear Energy Series publications address safety, it is ensured that the IAEA safety standards are referred to as the current boundary conditions for the application of nuclear technology.

In recent years, developments across various industries have illustrated the potential benefits of digital innovation. Many industries have improved the efficiency and economic performance of their facilities and products through the judicious use of advanced digital technologies while maintaining high levels of safety and security. The nuclear industry could greatly benefit from these experiences, in particular to improve the economic performance and competitiveness of existing and future nuclear power plants.

Many novel designs and advanced technologies are currently being explored for both existing nuclear power plants and new constructions, with the aim of establishing more efficient nuclear processes and improving design standardization. The amount of data acquired at nuclear power plants has gradually increased, and the ongoing development of data based, intelligent analysis technologies can support fault diagnosis and prediction for important equipment, facilitating the implementation of adequate operation and maintenance strategies. With the appropriate use of advanced instrumentation and control (I&C) systems and other advanced digital technologies, nuclear power plants can achieve highly efficient, competitive power generation in an increasingly complex operational context.

This publication reflects the importance of systems engineering in I&C systems for nuclear facilities as recognized by the IAEA Technical Working Group on Nuclear Power Plant Instrumentation and Control (TWG-NPPIC). It aims to assist Member States in understanding the judicious use of modern I&C systems and other advanced digital technologies to support optimal nuclear power plant performance and minimize operational costs while ensuring high levels of safety and security. The publication provides an overview of current knowledge, up-to-date best practices, experiences, benefits and related challenges, as well as the role of modern I&C systems and other advanced digital technologies in supporting the improvement and optimization of plant performance.

This publication was developed by a committee of international subject matter experts and advisors from Member States. The IAEA is grateful to all the contributors and reviewers listed at the end of this publication, in particular T. Nguyen (France), who served as chair of the authoring group. The IAEA officer responsible for this publication was N. Ngoy Kubelwa of the Division of Nuclear Power.

CONTENTS

1. INTRODUCTION

1.1. BACKGROUND

The instrumentation and control (I&C) of a nuclear power plant can be regarded as its 'nervous system', playing an important role not only in plant safety but also plant performance. This is one of the reasons for the pervasive use of digital technologies in nuclear power plant I&C nowadays, as they offer seemingly unlimited functional capabilities.

More generally, digital technologies beyond I&C — for example, modelling and simulation, virtual reality (VR) or augmented reality (AR) — may also be of great benefit to the performance of existing operational plants as well as of new builds. Performance improvement is often necessary for a nuclear power plant to remain economically competitive as the cost of wind and solar power decreases steadily.

In this publication, plant performance is interpreted in a very broad sense and is not limited to economic performance but also includes other factors such as environmental performance and flexible operation capabilities (in the face of increased ratios of intermittent electric power sources).

1.2. OBJECTIVE

The objective of this publication is to provide managers and other decision makers with an overview of how modern digital I&C systems and equipment, as well as other advanced digital technologies, may be used to ensure or improve plant performance. Because many new builds and operating plants already feature digital I&C systems, the focus is mainly, though not exclusively, on other advanced digital technologies that can help designers, construction workers, operators, maintenance staff and decommissioning staff be more efficient and make fewer errors.

Guidance and recommendations provided here in relation to identified good practices represent expert opinion but are not made on the basis of a consensus of all Member States.

1.3. SCOPE

As plant performance is an extremely wide subject, this publication does not try to describe the technical aspects of the suggested methods and technologies, nor how they are or may be implemented. Instead, it focuses on the services these methods and technologies can provide and how they can benefit plant performance. Where possible, short examples are given, preferably from the nuclear industry, but also from other industries when examples from nuclear power plants are lacking. Information on the level of maturity of these examples for the nuclear industry is also provided.

Readers interested in technical and implementation details can generally find them in referenced publications. Indeed, the IAEA, the International Electrotechnical Commission (IEC), the Electric Power Research Institute (EPRI) and many other research teams have extensively documented numerous aspects of, and approaches to, plant performance.

The introduction of any new method or technology has associated costs. As each nuclear power plant is a specific case, the development of business cases is out of the scope of this publication. Computer security needs to be taken into consideration whenever dealing with digital technologies and I&C. IAEA Safety Standards Series No. SSR-2/1 (Rev. 1), Safety of Nuclear Power Plants: Design [1], establishes the safety requirements for nuclear power plant design, and IAEA Safety Standards Series No. SSG-39, Design of Instrumentation and Control Systems for Nuclear Power Plants [2], provides the safety recommendations for the design of I&C systems. Therefore, this publication does not address the safety requirements or guidelines for the application of advanced digital technologies in nuclear power plants. More generally, this publication does not address the decision making process about the use or non-use of the presented methods and technologies (see Ref. [3], for example).

1.4. STRUCTURE

Besides the introduction and the conclusion, this publication is composed of six sections and seven annexes. Section 2 presents the main components of plant performance considered in this publication. Sections 3 and 4 each present a key generic digital technology for plant performance: modelling in Section 3 and improved diagnostics and prognostics in Section 4. Section 5 presents other means of improving plant performance. Section 6 presents generic technologies that could be used to improve plant performance. Lastly, Section 7 discusses some of the practical challenges related to the use of advanced digital technologies.

Each annex presents an approach or technique applied in some existing plants:

— Annex I presents Metroscope, a commercially available software application for real time diagnostics of industrial thermodynamic processes (Metroscope, France). These diagnostics are performed by leveraging real time operational data, past operating experience and artificial intelligence (AI).
— Annex II gives an overview of CADIS, a data driven decision support solution developed by Framatome (Germany), with selected application examples. With CADIS, condition, process and maintenance data are automatically collected and transformed into analytic insights for decision making and plant asset optimization.
— Annex III presents various advanced I&C and other digital solutions applied in Kraftwerk Union (KWU, Germany) nuclear power plants to support the performance of transients and maximize the flexibility capacity, taking into account ageing considerations. Examples are also included.
— Annex IV presents an on-line monitoring success story from the United Kingdom.
— Annex V illustrates how modern core physics simulation tools, digital twins and new digital based methodologies increase the efficiency and performances of nuclear industry processes for the benefit of the end users, with a selection of practical examples from Tractebel Engineering (Tractebel, Belgium).
— Annex VI presents various digital solutions applied by CEZ Group (Czech Republic) for process optimization. These solutions support plant personnel in safe and effective operation.
— Annex VII presents applications for plant performance enhancement using virtualized digital I&C in Generation III+ nuclear power plants in the Republic of Korea.

2. PLANT PERFORMANCE OVERVIEW

This publication considers six top level, complementary components of plant performance:

— Plant availability;
— Efficiency of engineering for design and construction;
— Optimization of operation and maintenance;
— Operational flexibility;
— Cogeneration;
— Environmental performance.

With nuclear power plants, the costs of electricity production are relatively stable compared to coal and gas plants. The main reason is that, for nuclear power plants, fuel represents only a small fraction of the total expenses: a doubling of the cost of uranium increases the cost of electricity by only about 10% [4]. However, as the costs of wind and solar power are steadily decreasing, an important objective of I&C and advanced digital technologies is to contribute to nuclear power plant cost reduction.

Nuclear power plant costs can be divided into capital costs and plant operation costs. Capital costs consist largely of design and construction expenses, including the substantial financial burdens associated with long design and construction times. Sections 3 and 6 present advanced digital technologies that can contribute to reducing these costs.

Plant operation costs can be further divided into fuel costs, operation and maintenance costs, waste disposal costs and decommissioning costs. Operation, maintenance and decommissioning costs include personnel costs and some material costs. Some of the technologies presented in Sections 3–6 may help reduce plant operation costs.

When striving to improve plant performance, it is generally preferable to make decisions and assess progress based on key performance indicators. These indicators need to be selected and defined with great care, to ensure that they are indeed representative of the desired performance objectives and that they are based on available, objective and accurate measurements. When defining key performance indicators, it is also important to consider that performance factors are sometimes antagonistic: improvement of one may come at the expense of another.

2.1. PLANT AVAILABILITY

Different approaches contribute to maximizing plant availability (i.e. the plant's ability to produce the desired level of power at any given instant).

2.1.1. Minimization of unplanned outages and unplanned power reduction

Various causes may lead to unplanned outages or undesired power reduction, including design faults, successful malicious attacks, plant equipment failures, instrument miscalibration or human error. Extensive guidance is already available to minimize the potential for design faults (for I&C, see for example Refs [2, 5–11]), to provide adequate protection against malicious attacks (see for example Refs [12–14]) and to ensure adequate training for personnel (see for example Ref. [15]). To further reduce the potential for unplanned outages and power reduction, the following techniques can also be applied:

— Improved diagnostics and prognostics (see Section 4);
— Minimization of human error in the control room with adequate consideration of human factors engineering (HFE) in the design of I&C system–human interfaces (see Ref. [16] and Section 6.7);
— Minimization of human error in the field during maintenance and outages (see Sections 5.6, 5.8 and 6.8).

2.1.2. Optimization of the duration of planned outages

The time interval between two planned outages mainly depends on plant architecture and process design, as do most of the tasks to be performed during an outage. However, efficient condition monitoring and accurate, dependable prognostics (see Section 4.1.1) may be used to better plan the necessary maintenance activities of an outage. Adequate support for surveillance testing and diagnostics and judicious mobile digital assistance (see Section 6.8) may improve effectiveness in the implementation of tasks. Advanced digital technologies may be useful for optimizing scheduling, given the very high number of tasks to be performed, the many constraints between tasks, and the fact that unforeseen tasks or constraints often arise.

2.1.3. Plant life extension

Plant life extension involves extensive analyses, including assessment of the residual life of plant components and plans for equipment replacement or refurbishment when necessary. In this context, I&C may be used as follows:

— To monitor the factors that influence the lifetime of components and passive structures essential to plant lifetime in order to estimate their remaining lifetime as accurately as possible;
— To control or help control these factors, when possible, to avoid unnecessary shortening of the lifetime of these components and structures and that of the plant itself (see Ref. [17]).

Another possibility is to use advanced methods to estimate the residual lifespan of plant components based on the environmental conditions to which they have been exposed (see Refs [18–20]). As I&C systems and equipment (including cabling) may themselves age and become obsolete, measures can also be taken to ensure that they can be replaced when necessary (see Ref. [21]).

2.2. EFFICIENCY OF ENGINEERING FOR DESIGN AND CONSTRUCTION

Nuclear power plants need long design and construction times incurring high capital and financial costs, which have large impacts on plant economic performance. Advanced digital technologies may help to improve the efficiency of design processes by:

— Optimizing the cost and duration of design and construction, in particular by the application of advanced systems engineering approaches (see Ref. [22]), modelling and simulation (see Section 3);
— Achieving a safe, secure and dependable plant design and an efficient, optimized operation.

Systems engineering approaches, modelling and simulation may also be used to support innovation and the introduction of features such as those proposed in Section 5.

2.3. OPTIMIZATION OF OPERATION AND MAINTENANCE

Digital I&C and advanced digital technologies may be used to provide operational aids to plant personnel to help them operate and maintain the plant and its systems and equipment as efficiently as possible (see Sections 3–5).

2.4. OPERATIONAL FLEXIBILITY

Most nuclear power plants already in operation are optimized for baseload operation where frequency stabilization and grid balancing are performed by other power generation means (typically hydroelectric, gas fired or coal fired plants). However, some Member States have extensive experience in flexible operation where nuclear power plants contribute to both frequency stabilization and grid balancing. Also, increased ratios of intermittent sources in energy mixes and more frequent events where power production exceeds demand have led other Member States to initiate feasibility studies and pilot projects for flexible operation. The IAEA describes flexible operation modes [23] including:

— Frequency regulation (often or continuous);
— Planned load following, including extended low power operation;
— Unplanned load following.

Flexible operation with its grid service capabilities may provide additional revenue. It could also improve fuel management and increase the time period between outages.

Adaptation of reactor power can be obtained by different means. For example, in a pressurized water reactor (PWR), control rods or boration/dilution may be used. These measures can also be combined. Each strategy leads to certain operational costs, including expenditure of the rod drives, burnup of the control rods, costs for boric acid and demineralized water, and effluents disposal. Moreover, flexible operation changes the power density distribution in the reactor core and influences the burnup distribution. A non-uniform burnup reduces the integral energy yield of fuel elements. Another factor to take into consideration is xenon poisoning, because frequent power adaptation could lead to an imbalance in xenon distribution. The complex xenon dynamics could lead to spatial power oscillations in the core, so-called xenon oscillations.

In water cooled reactors, it is also necessary to address issues induced by pellet–cladding interaction (PCI) (see Ref. [24] and Section 3.10.2(a) for an implementation within a limitation system).

Even for experienced human operators, it is difficult to optimize the overall costs of flexible operation, due to the many factors that have to be taken into account. Operational aids can, however, make predictions on the state of the core (e.g. for the next 24 hours based on a given load schedule, current rod position and core poisoning). Such aids can propose control actions avoiding xenon oscillations, ensuring compliance with the load schedule, saving the control resources mentioned above and optimizing operation costs.

2.4.1. Benefits from a utility viewpoint

Grid codes define mandatory requirements for grid stabilization and balancing, thereby establishing the overall framework for flexibility of production means. In particular, grid code evolutions may impose new mandatory flexibility requirements on the production methods already in operation, including nuclear power plants.

I&C may be used to introduce, enhance or support the flexibility of a nuclear power plant in order to meet grid code requirements and/or increase profitability. Flexibility enables so-called economic dispatch, by which a fleet-operating utility minimizes the total cost of meeting electricity demand using a judicious selection of production methods, possibly with the help of digital aids, taking into account the flexibility, state and costs of its production methods (see Section 5.2). Flexibility of each method, in particular of each nuclear power plant, provides more optimization options and improves profitability at fleet level. In this context, digital aids may also help determine the best moment for restoring full power for a given nuclear power plant and the most efficient manner to do so.

Flexibility may also contribute to outage optimization by planning the unavailability of nuclear power plants for times when the demand and market prices are at their lowest, also taking into account the availability of outage teams.

Some of the well established solutions for flexible operation also help improve reliability and safety, for example by introducing automated core control functions or by helping operators managing reactivity (see Sections 3.7 and 3.10 and Annex III).

2.4.2. Benefits from a grid viewpoint

Grid stabilization and balancing are increasingly critical, in particular due to increasing ratios of intermittent sources in energy mixes and also to their lack of inertia. Nuclear power plants' flexibility increases the ability of grid dispatchers to stabilize and balance the grid. For example, they can more easily transmit power from renewable sources to regions with high demands, as can be often seen in Germany, where most of the offshore wind parks are in the north and high demands from industry are in the south. The overall grid stabilization and balancing costs can also be better optimized with more numerous flexible sources.

2.4.3. Examples of features supporting flexibility

The following are some illustrative examples of features supporting flexibility:

(a) Primary and secondary frequency control in turbine I&C: Different plant systems contribute to flexibility. One of them is the turbine I&C when it provides primary and secondary frequency control functions. If necessary, these functions may be added through an upgrade of the turbine I&C (see Annex III).

(b) Enhanced reactor I&C: Reactor I&C can implement functions mitigating the adverse effects of flexible operation and ensures the thermal power needed by the turbine (see annex I–4.5 of Ref. [23]). With modern digital I&C, improved power distribution and predictive reactivity management enable profitable ancillary and balancing services in a large power band, optimize fuel management and reduce operator burden in rapid transients (see Annex III).

(c) Advanced digital tools: Ageing effects like fatigue, vibration or flow induced corrosion also need to be taken into account. Inspection intervals and maintenance programmes can be optimized based on enhanced modelling and simulations, including big data analyses, without compromising safety and availability (see Annex III).

2.4.4. Maturity

Experience gained over many decades in some Member States shows that nuclear power plants equipped with appropriate measures can be more flexible than coal/oil plants and can be comparable to combined gas plants [25]. However, flexible operation can lead to increased wear and tear of process equipment, increased maintenance costs and longer outages. Optimized monitoring and condition based or predictive maintenance can reduce the additional costs and events due to flexible operation.

2.5. COGENERATION

For nuclear power plants, as for any thermal power plant, the efficiency of electricity production is limited by the second law of thermodynamics and Carnot's theorem. For example, with a typical PWR, around two thirds of the generated heat cannot be converted into electricity. Cogeneration is the integration of electricity production with other applications so that heat that otherwise would be wasted and pollute the environment can serve other purposes, for example, urban heating, industrial processes, desalination or hydrogen production (see for example Refs [26–28]).

Cogeneration options depend on temperature, and Fig. 1 shows possible options. Temperature itself depends on reactor type, and Fig. 2 shows the typical outlet temperature of different types of reactors. The temperature available for cogeneration is the one at the steam or water extraction point.

2.5.1. Benefits to plant performance

Cogeneration can provide many economic and environmental benefits. Economic benefits result from getting value from heat that would otherwise be wasted. They may also result, with some of the cogeneration options, from the ability to operate at full power even when electricity demand is low. However, it is important to consider the need for additional systems to support cogeneration, as these systems enable the simultaneous production of electricity and heat. One of the key challenges in cogeneration is multiload coordination, which involves managing and balancing the varying demands of electrical and thermal loads. This complexity requires careful integration of control strategies and system components to ensure efficient operation and responsiveness to dynamic load conditions. Also, not all cogeneration options can make use, at all times, of all the heat unused by electricity production. For example, demand for urban heating is much lower during summer.

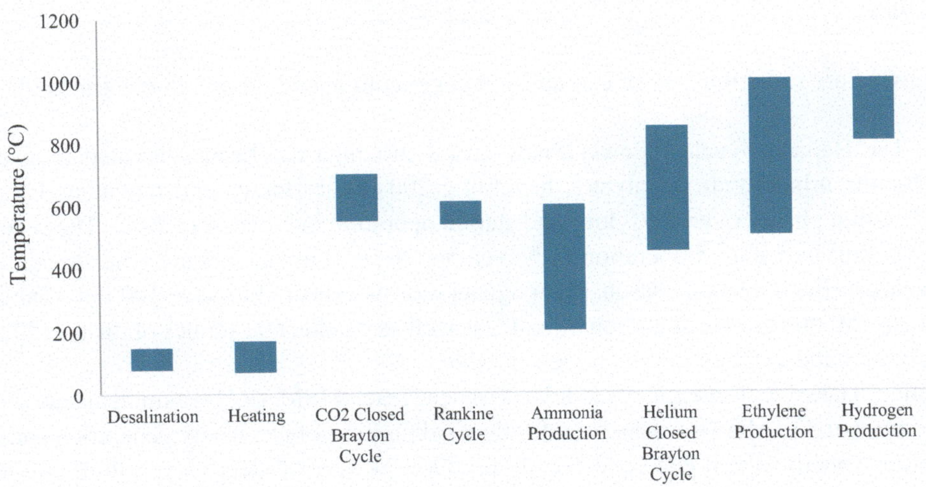

FIG. 1. Typical cogeneration applications based on outlet temperature (°C) (courtesy of F. Song, Shanghai Nuclear Engineering Research and Design Institute, China).

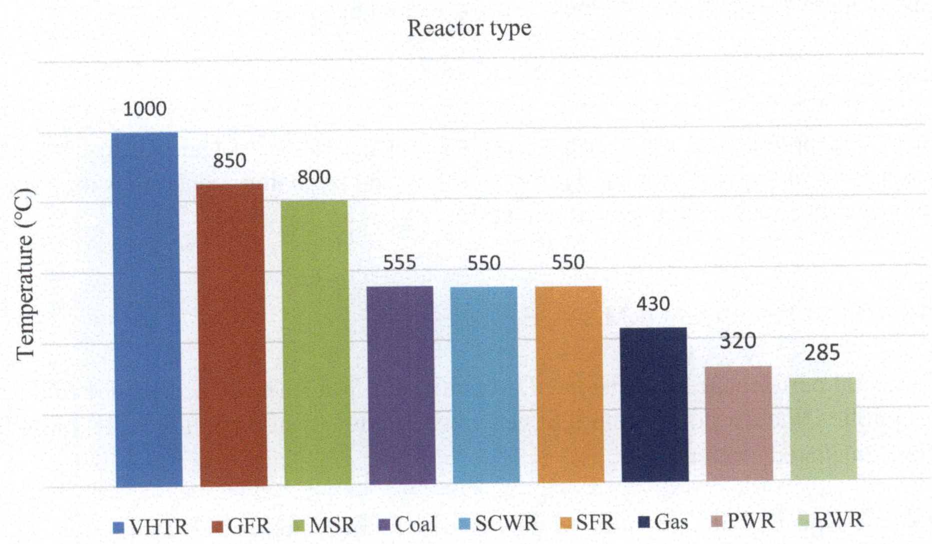

FIG. 2. Typical outlet temperature of different types of nuclear reactors. BWR — boiling water reactor; GFR — gas cooled fast reactor; MSR — molten salt reactor; PWR — pressurized water reactor; SCWR — supercritical water reactor; SFR — sodium cooled fast reactor; VHTR — very high temperature reactor. (Courtesy of F. Song, Shanghai Nuclear Engineering Research and Design Institute, China.)

Environmental benefits result from the lower amount of wasted heat that needs to be dispersed in the environment. They also result from the fact that nuclear power can replace more polluting processes, as shown in the examples below.

2.5.2. Examples

The following are some illustrative examples of cogeneration and other non-electric applications:

(a) Heating: The Haiyang Nuclear Power Plant Unit 1 in China has been renovated to also provide heating for the urban area of Haiyang, benefiting 200 000 residents and replacing 12 coal fired boilers. The first phase covered 0.7 km^2 and started operation in November 2019. The second phase covered 4.5 km^2 and started operation in November 2021. Haiyang has now bid farewell to years of coal heating. Each heating season, this cogeneration is expected to save 100 000 tonnes of coal and to avoid 180 000 tonnes of carbon dioxide, as well as smoke, 691 tonnes of dust, 1123 tonnes of nitrogen oxides and 1188 tonnes of sulphur dioxide.
(b) Desalination: The system-integrated modular advanced reactor (SMART) is a small reactor with a rated thermal power of 330 MW. It can be used for the dual purpose of electricity generation and seawater desalination. Specifically, it can concurrently produce 90 MW of electricity and 40 000 m^3/day of desalinated water, which is sufficient for 100 000 residents [29]. Nuclear desalination is generally very cost effective compared to fossil fuels [30].
(c) Medical isotopes production: In recent years, the production of medical isotopes in nuclear power plants has become profitable, since demand for radiopharmaceuticals is on the rise and, at the same time, the number of research reactors is decreasing [31]. One of the best known isotopes in this context is lutetium-177. With full power conditions, its production time is around 14 days and its half-life is of only a few days. Thus, it needs to be produced continually. With a rather small investment, up to 1.5 kg of the isotope can be generated per year.

2.5.3. Maturity

Up to now, cogeneration has not been a widespread practice in the nuclear industry. It is a technically mature approach with increasing demand. However, safety and regulatory constraints may apply, and not all sites are eligible for a particular cogeneration option.

2.6. ENVIRONMENTAL PERFORMANCE

Environmental performance is considered, in particular, to further the acceptance of nuclear power in the general public. Nuclear power naturally generates low levels of carbon dioxide, but other aspects also have to be considered, such as:

— Radiological aspects, with the need to generate as limited amounts of radioactive effluents and waste as possible during normal operation. (As part of safety, the prevention of accidents and the minimization of their radiological consequences are not considered in this publication.) Innovations in process design could be facilitated by the application of systems engineering approaches (see Ref. [22]) and modelling and simulation (see Section 3). Advanced radiological services (see Section 3.12) could be used not only to improve radiological performance, but also as a real time help to, and an efficiency booster of, field operators and outage personnel.
— Thermal aspects, with the need to avoid elevating the temperature of the cold source beyond authorized limits. To go beyond what is generally achieved in nuclear power plants, a possible approach is to use heat that would otherwise be wasted for purposes other than electricity production (see Section 2.5).

2.6.1. Examples

The following are some illustrative examples of technological and operational measures for improving environmental performance:

(a) Minimization of effluents:
  (i) The use of boron, particularly for normal operation, is a significant factor in effluents production. Since the total amount of coolant in a water cooled reactor remains nearly constant, the quantity of boron related effluents is equal to the quantity of injected boric acid and demineralized water. Flexible operation may increase the consumption of boric acid and demineralized water, and, consequently, the volume of effluents. Core monitoring systems (CMSs) (see Section 3.3.2(e) and Annex V–3.1) and advanced control algorithms (Section 3.10 and Annex III) can help to minimize boron consumption. See also Refs [32, 33] for AI based approaches.
  (ii) More radically, advanced digital technologies, in particular modelling and simulation, may be used to enable and facilitate plant process innovations aimed at reducing the volume of effluents in general and at limiting or avoiding altogether the use of boron for normal operation.

(b) Minimization of fuel damage: In water cooled reactors, fretting and corrosion are significant causes for fuel damage, but PCI also plays an important role, when local changes in power density cause thermal expansion of fuel pellets. Digital I&C can prevent fuel damage, for example by comparing measured power densities provided by in-core instrumentation with a power limit value calculated using a simplified fuel conditioning model (see Section 3.10, Annex II and Ref. [24]).

(c) Advanced real time dispersion modelling: Advanced modelling and faster-than-real-time simulation may be used to reduce the radiological impacts of radioactive releases into the environment, both during normal operation and after accidents. Mitigating measures can be better informed in real time with recent, powerful simulation means and model developments specifically aimed at improving near field dispersion analyses.

2.6.2. Maturity

Various works worldwide improve nuclear power plants' environmental performance. The wide spectrum of available examples, from efforts to reduce fuel damage to AI applications in advanced control concepts, highlights the growing importance and understanding of the topic.

3. MODEL AIDED SYSTEMS ENGINEERING

Model aided systems engineering is an approach that integrates modelling techniques into the systems engineering process to enhance the design, analysis and management of complex systems. By leveraging digital models, model aided systems engineering supports better decision making, improves traceability and facilitates communication among stakeholders throughout the system life cycle.

3.1. MAIN CONCEPTS

This methodology enables the representation of system requirements, architecture, behaviour and constraints in a structured and visual format, promoting consistency and reducing ambiguity. Model aided systems engineering also supports early validation and verification activities, risk assessment and change impact analysis, contributing to more robust and adaptable system designs.

3.1.1. Models

A model is a selective and simplified representation of an actual, real life system, entity, phenomenon or process. George Box (1919–2013), a British statistician, had this famous aphorism: "All models are wrong, but some are useful" [34].

A digital model is a model expressed in a format with well defined syntax and semantics, so that it can be extensively analysed and exploited with software tools. A semidigital model is a model expressed in a format mixing parts with well defined syntax and semantics with parts relying on natural language and informal diagrams: software tools can fully support the digital parts but can provide only very limited support for the non-digital ones. In this publication, except when explicitly specified otherwise, the term 'model' stands for 'digital model' or 'digital parts of a semidigital model'.

Also, this publication uses the expression 'model aided systems engineering' in lieu of 'model based systems engineering', which is more commonly used in the systems engineering literature but that generally makes very limited use of simulation or analysis; to master complexity and ensure engineering efficiency, these techniques are essential, and 'model aided systems engineering' is a shortcut for 'model-, simulation- and analysis-aided systems engineering'.

There are many different types of models:

— Behavioural models express dynamic phenomena (i.e. phenomena that occur over time). A wide variety of such models may be used to the benefit of plant performance:
  - Behavioural requirements models specify expected behaviours and may be used to automatically check that an observed, simulated or computed behaviour is consistent with specified requirements, opening the way to early and rigorous testing of designs and implementations.
  - Functional models are a particular type of behavioural requirements models specifying expected behaviours in terms of functions to be performed, in a structured and hierarchical manner.
  - Probabilistic models express how probabilistic safety or availability aspects are related to the failure probabilities of individual components and possibly also to operating conditions, probabilities of human error and potentials for systematic failure. Probabilistic models are also widely used in radiation physics and engineering.
  - Economic models may be used to estimate different cost and revenue aspects, possibly taking into account operation and market assumptions, opening the way to economic forecasting and optimization.
  - Task scheduling models identify the tasks necessary to achieve a given goal and represent the scheduling constraints applicable to those tasks, opening the way to tool supported schedule optimization.
  - Process, physics and multiphysics models represent physical phenomena (e.g. thermodynamic, electric or mechanical phenomena), opening the way to a better understanding of plant process behaviour in various conditions and facilitating the study of innovative process designs.
  - Steady state models are often simpler models that can be simulated in real time or faster-than-real-time during operation, providing many benefits to plant performance, including process optimization, performance tracking, early fault detection and identification (see Annex I) and precise calculation of unmeasured quantities (see Annex V).
— Geometric (3-D or 2-D) models may be used for a wide variety of activities beneficial to plant performance, such as:
  - Design of the shapes and relative positions of objects and structures;
  - Verification of non-collision (i.e. that two or more objects or structures do not geometrically encroach upon one another);
  - Preparation of manufacturing, construction and installation on-site;
  - Verification that manufacturing, construction and installation on-site are indeed within acceptable geometric tolerance to design.
— Kinematics models are both behavioural and geometric models, expressing how shapes, sizes and relative positions evolve along time.
— Calculation models specify how a quantity of interest can be derived from other known quantities.

Models can vary significantly in their level of detail. In the early stages of the engineering process, when system specifications are still being defined, models tend to be coarse, capturing only broad constraints and expected envelopes for shapes or behaviours. As the design matures, these models are progressively refined to represent systems and components with greater precision, enabling more thorough verification, analysis and optimization.

Despite their simplicity, coarse models offer substantial value to plant performance. They facilitate the early application of advanced software tools, supporting efficient engineering workflows, preliminary design verification and initial design optimization. This early modelling capability helps identify potential issues sooner and supports informed decision making throughout the system development life cycle.

The term 'digital twin' is often used nowadays to designate a model serving as a digital counterpart of a physical object or system, as it was or will be built and installed on-site (see Annexes I, V and VII). However, for systems as complex as nuclear power plants and their associated systems, including nuclear power plant I&C systems, no single model can represent all aspects of interest at the necessary level of detail. Thus, they can have many different digital twins, each serving different purposes and used for different engineering activities, at different levels of detail or abstraction.

Modelling is a very well established technique in all industrial sectors, including in the nuclear industry. However, it is a protean technique covering an extremely wide variety of aspects and many disciplines, and its use in the nuclear industry is often not as widespread and systematic as it could be. For more detailed information on systems engineering and modelling, see Refs [22, 35, 36].

As shown in the following subsections, models may be used to support all the stages of a nuclear power plant's life cycle, from initial concept studies through to operation and then decommissioning. Some, or even a large part, of the models that supported design and construction can be used to support operation and decommissioning. In the case of existing plants, models may have played a limited role during design and construction and may need to be created, as shown in Annex I.

3.1.2. Simulation

Simulation is a process in which the behaviour that has been specified by a behavioural model over time is made explicit and is given initial and boundary conditions. In the case of a non-deterministic behavioural model, this would be 'a' behaviour (i.e. unspecified). One can say that simulation 'gives life to' and 'animates' a model. Comprehending the full implications of a model is essential for both its authors and users. Even relatively simple models can be challenging to interpret manually without the support of simulation tools. As models begin to describe complex systems in greater detail, manual analysis becomes increasingly difficult — and in many cases, practically impossible. Simulation enables a deeper understanding of system behaviour, interactions and constraints, making it an indispensable tool in model aided systems engineering. If the model is a correct representation of a real world system or phenomenon, then simulation helps predict how that system or phenomenon will evolve in time.

With appropriate models, simulation may be used for many different purposes beneficial to plant performance, for example:

— To improve the efficiency of individual teams (see Section 3.14).
— To improve the cooperation between and the coordination of teams, disciplines, organizations and stakeholders by:
  - Helping them, using co-simulation (i.e. synchronized simultaneous simulation of multiple models representing their different viewpoints), realize the effects on their own work of decisions made by others.
  - Enabling an agile and effective concurrent engineering workflow (see Section 3.15 and Annex VI).
— To ensure, early in engineering processes, that:
  - Specified behavioural requirements (i.e. functional and performance requirements) are indeed in favour of plant performance and are free from errors that could be detrimental to performance, safety and/or security.
  - Contemplated engineered solutions do satisfy the specified behavioural requirements.

- Errors are discovered early — since those that remain until late in the plant life cycle are often very costly not only in purely monetary terms but also in terms of delays, or even outright catastrophic in terms of safety or security (see Section 3.2).
— To help find more optimal solutions (see Section 3.3) and to innovate while guaranteeing safety and security (see Section 3.4).
— To mimic system responses in training simulators (see Section 3.5), possibly in VR or AR environments (see Section 6.9).

In addition to these direct benefits to plant performance, simulation has more indirect benefits such as providing technical support for the maintenance of plant specific engineering, safety and security knowledge for the lifetime of the plant (see Section 3.13).

Simulation is a very mature technique that is widely applied in all industrial sectors, with proven benefits in terms of engineering efficiency, design and operations optimization. Although it is also applied in the nuclear industry, its usage could be much wider and more systematic.

3.1.3. Analysis

Model analysis aims at calculating a feature of interest of a model or at proving or disproving, possibly under specified external conditions, that a model satisfies a certain required property. Different techniques may be applied, depending on the nature of the model and of the feature of interest or required property, but all need to be based on rigorous and well founded principles. For example:

— Analysis of geometric computer aided design (CAD) models may be used to calculate features such as volume, mass or position of gravity centre and to prove or disprove structural properties such as solidity or deformation under mechanical stress.
— Analysis of probabilistic models may be used to estimate system level failure rates or probabilities, based on the failure rates of individual components for various failure modes. These techniques are very mature and widely applied in the nuclear industry.
— Analysis of process models together with real time operations data may be used to improve operational efficiency, for example by:
    - Reducing measurements' uncertainty margins and discarding grossly incorrect measurements (see Sections 3.6 and 3.7);
    - Providing support to risk informed operation (see Section 3.8);
    - Helping optimize fleet assets management (see Section 3.9);
    - Helping to determine appropriate courses of action (see Section 5.3);
    - Supporting advanced radiological services (see Section 3.12).
— Steady state analyses may be used to determine optimal set points during normal, stable operating periods.

To prove or disprove a property, analysis techniques such as the following can be used:

— Model checking, which systematically examines the model under all specified conditions. If under some of these conditions the model does not satisfy the property, one benefit of the technique is that it can exhibit a counter example.
— Theorem proving, which applies step-by-step logical reasoning to demonstrate that the property is a logical consequence of the model and of the postulated external conditions.

In addition to these direct benefits to plant performance, model based proof of properties has more indirect benefits such as attracting and retaining talent.

Proof of behavioural properties (e.g. properties that depend on time), though extremely useful, has up to now been applied only in a limited number of industrial applications. In the nuclear industry, it is applied in some Member States as an independent confidence building measure in the verification of safety digital I&C systems.

3.1.4. Information retrieval

In the engineering of systems as large, complex and critical as nuclear power plants and their associated systems, including nuclear power plant I&C systems, the volume of information that needs to be mastered is staggering, and the ability to retrieve desired pieces of information is of the utmost importance. When that information is essentially provided in traditional documentation (i.e. in natural language and informal diagrams), such retrieval is often difficult and time consuming and sometimes unsuccessful: up to now, no current software tool can provide for engineering the information retrieval ability of some Internet search engines. However, as models bring structure and well defined syntax and semantics, information retrieval tools can be more efficient.

A huge volume of data is produced during the operation of a nuclear power plant (by sensing devices but not only). Such 'big data' datasets contain numerous pieces of information that could be extremely useful to plant performance, but extracting this information is often a challenge and is the subject of a new and very active field of research (see Section 6.13).

3.2. VERIFICATION OF SYSTEM OPERABILITY

The situations a system may face result from the combination of external conditions, internal states and operational goals assigned to the system. Some conditions and states may be discrete (e.g. a switch may be opened or closed), others may be continuous (e.g. a temperature may vary continuously within a certain range). Operational goals may vary in time (e.g. stable power production, shutting down). To verify that a system indeed satisfies a requirement under all specified situations, it is necessary to examine and test the system in all of them.

This is, in general, infeasible with an actual physical system, due to constraints related to time (there are too many possible situations, and the examination of each takes significant time), safety (some situations such as accidents are too dangerous or may lead to unacceptable consequences) and cost (some situations would require costly equipment or cause costly damage to the system). A possible alternative approach is to use a model as a proxy of the actual system and simulation to explore the domain of possible situations. Even when physical testing is possible, simulation is generally cheaper and faster and thus allows the examination of a larger number of situations. Also, simulation can be done earlier in the engineering process than physical testing, as shown by the example in Section 3.15.2.3, and it may be used to verify not only the detailed design of the system but also the adequacy of its requirements and the correctness of each design step.

Traditional simulation methods imply manual definition of each simulated scenario and manual verification so that in each case no requirement is violated. Recent simulation methods and tools allow automatic generation of test cases and automatic verification of compliance with requirements, opening the way to massive simulation, where extremely large numbers of scenarios are examined (in the thousands, tens of thousands or even more). With such automation, the simulation process may be repeated at limited cost and effort whenever changes are made in the model, in the scenarios or in the requirements.

In favourable cases, formal analysis can be applied to systematically explore all specified situations. There are, unfortunately, theoretical and practical limits to what can be done through formal analysis.

3.2.1. Benefits to plant performance

Besides a more efficient, more streamlined and less costly design process, simulation and most particularly automated massive simulation may be used to reduce the risk of overlooking significant situations, or that some specified requirement might be inadequate, or that the solution might not comply with some requirements. Such errors could cause delays during design and construction and loss of production or even severe accidents during operation.

3.2.2. Examples

The following are some illustrative examples of model aided systems engineering approaches to the verification of system operability:

(a) Automated failure mode, effects and criticality analysis (FMECA) of a heating, ventilation and air conditioning (HVAC) system: FMECA is a notoriously time consuming and expensive technique, where the effects of each possible failure mode of each system component are analysed to determine whether they are acceptable. For real life systems, it often implies hundreds or thousands of different cases. When performed manually, there is an inherent risk of error in the analysis. Modelling and automated simulation were used in a new build design to systematically examine all single component failure scenarios of a safety HVAC system and to fully automate the FMECA process. Each time a detailed design was modified, FMECA could be repeated with minimal effort, cost, risk of error and delay.

(b) Priority logic verification in safety I&C: Any inadequacy in the functional requirements of a priority logic system could lead to spurious actuation (with adverse effects on plant availability) or could annihilate defence in depth and diversity efforts and lead to failure to actuate (with possibly disastrous consequences for plant integrity). In the design of the safety I&C of a plant, simulation was used to verify their adequacy in all specified situations. Several errors were found and corrected before proceeding to manufacturing.

(c) Cranbrook manoeuvre reconstruction: The Cranbrook manoeuvre is an aircraft accident that occurred in 1978, killing 42 people. It was in part caused by a thrust reverser requirement that was adequate in most situations but was woefully inappropriate in the situation of the accident. With the exploration of scenarios with modern modelling and automated simulation techniques and tools, the accident scenario was revealed in a matter of minutes (with one hour of work to develop the model).

3.2.3. Maturity

Modelling and simulation are widely applied in all industries, including the nuclear industry. However, in most cases, they are applied manually to very specific and local design aspects. Methods and tools enabling automated massive simulation have appeared only recently but are now commercially available and applied by some in the nuclear industry. They were applied in all three of the examples described above.

3.3. OPTIMIZATION OF SYSTEMS DESIGN AND OPERATION

When different engineering teams, disciplines and organizations work separately with limited technical coordination, each tends to take less-than-optimal margins to ensure that the whole can indeed operate as required. This may lead to overdesigned, more complex and more costly solutions than strictly necessary and inefficient plant processes.

Margins are also taken when system behaviour is difficult to calculate or estimate manually, to ensure that operation will not exceed authorized limits. This may lead to the system operating in suboptimal conditions.

3.3.1. Benefits to plant performance

The Functional Mock-up Interface (FMI) industrial standard (see Ref. [37]) allows co-simulation, where models for subsystems developed by different parties can be simulated together so that it is possible to verify whether the functional and performance requirements of the overall system are satisfied even with optimized margins and simplified designs.

Modelling, simulation and process data reconciliation (see Section 3.6) and assimilation (see Section 3.7) can also be used to provide better understanding and estimates of the possible extent of system behaviour and to operate closer to optimal conditions.

When the operation of a system has many degrees of freedom, it is sometimes difficult to determine the most optimal control setting. Modelling, together with simulation or specific optimization techniques, may be used in finding a course of action close to that optimum.

Lastly, with modelling and simulation, it is possible, at limited cost and with moderate effort, explore multiple design or operational solutions, discard those that do not meet requirements and determine which of the remaining ones are the most optimal. This can be done manually to explore very different solutions, or with advanced techniques based on genetic algorithms or AI.

Modelling and simulation, and sometimes formal analysis, may be used to reduce design, construction and maintenance costs; to reduce effluent production and resource consumption; and to improve power production and flexibility.

3.3.2. Examples

The following are some illustrative examples of model aided systems engineering approaches that support the optimization of systems design and operation:

(a) HVAC optimization: Automated FMECA, as presented in Section 3.2.2, was applied to a safety HVAC system, using a thermodynamic model of the 'clients' of the system. Under some of the failure conditions, the analysis showed that the system was unable to provide sufficient cooling power. This would normally have led to a costly design modification. However, the thermodynamic model also showed that the 'clients' would exceed their maximum temperature only after a delay that was amply sufficient for repairing the HVAC system. Thus, no modification was deemed necessary.

(b) Plant uprating: Modelling, simulation and analyses may be used to prepare the uprating of an operating plant, for example by providing better estimates of:
    (i) Measurement uncertainties (see for example Section 3.6);
    (ii) The effects of modified process parameters (see for example Section 3.11.2.2);
    (iii) The effects of new or upgraded and more efficient plant components and systems.

(c) Modelling, simulation and analyses may naturally assist the design of the uprate and the training of personnel. They may also contribute to the safety justification, when some parameters, components and systems are important to safety.

(d) Spent fuel burnup credit: Burnup credit criticality safety analysis determines the minimum required burnup for fuel assemblies to be stored in the spent fuel pool or loaded into transport casks. A new, more accurate Bayesian method was used to justify a reduction of fuel assemblies' storage distance. Even after consideration of the computational and experimental uncertainties, this resulted in an increase of the storage capacity of the spent fuel pool on the order of 20% for a PWR, as illustrated in Fig. 3, and overall, in a very significant economic benefit [38].

(e) Core monitoring and operational support: Through calculations and comparisons with on-line measurements, CMSs in combination with advanced 3-D core simulators can provide extensive real time insights into the state of the core and, therefore, support flexible and low cost operation while maintaining the highest level of safety. CMSs have accumulated more than 120 years of operating experience, are available for most types of light water reactors, different types of measurements (e.g. ex-core/in-core measurements, aero-ball measurements or movable flux chamber systems) and can be customized to specific plant configurations. State of the art CMSs feature a broad spectrum of functionalities, such as:

(i)   Tracing, at each point in time, critical conditions (e.g. shutdown margins), limiting power peaking factors and other technical specification limits;
(ii)   Accurate reconstruction of 3-D power distribution, core parameter trending and monitoring of thermal limits;
(iii) Integrated flux mapping (with and without xenon equilibrium) [39];
(iv)   Burnup and isotopic tracking;
(v)   Ergonomic, graphical and configurable reporting to, and interfaces with, operators, to reduce the potential for human error (see Section 6.7.2.6);
(vi)   High levels of automation to facilitate operation and reduce operating costs;
(vii) Predictive capabilities to support flexible operation with, for instance, an easy planning of both startups and power manoeuvrings for efficient load following, reduced boron consumption and reduced downtimes due to xenon effects;
(viii) Facilitated cause analyses of unusual effects on fuel assemblies.

3.3.3. Maturity

Use of modelling and simulation to optimize specific and local design aspects is a widespread practice in all industries, but in the nuclear industry, it is not systematically applied to all aspects that could benefit from it. The use of modelling to optimize the steering of an installation and of co-simulation for global design optimization is relatively recent but is well supported by commercially available tools.

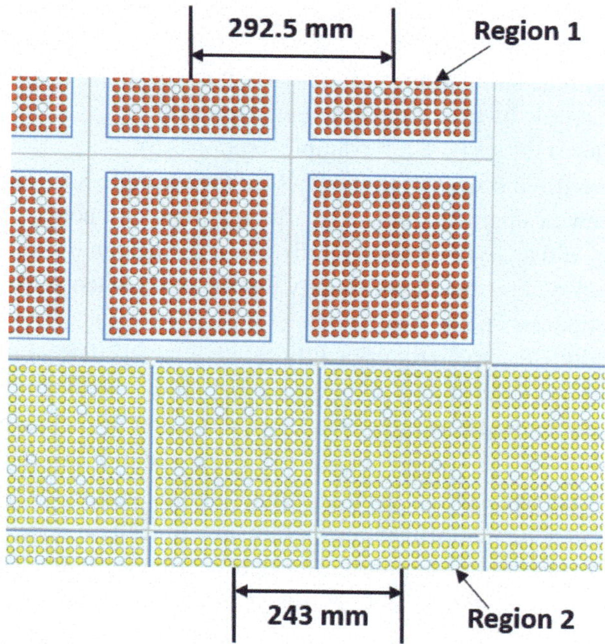

FIG. 3. Horizontal section of a storage rack arrangement in the spent fuel pool, showing standard spacing (Region 1) and compacted spacing (Region 2) (courtesy of Framatome, Germany).

3.4. ENABLING PLANT INNOVATION INITIATIVES

The nuclear industry often proceeds with step-by-step evolutions, where the design of a new plant or series introduces limited changes to a preceding, proven-in-use design; with safety being the most prominent consideration, radical innovations tend to be left aside. However, in the face of increasingly cost effective non-nuclear sources of power, performance is now an equally essential consideration. Recent designs, in particular pressurized water small modular reactor (SMR) designs, introduce more numerous and sometimes radical innovations: mutualized operation of multiple reactors (see Section 5.1), extensive use of passive safety and non-safety features and systems, integrated primary circuit designs (where the primary circuit is completely integrated within the reactor vessel), boron-less normal operation, extended fuel cycles, etc. Generation IV designs, by definition, introduce even more radical innovations.

3.4.1. Benefits to plant performance

Modelling, simulation and analysis may help verify, from very early engineering stages, that innovations indeed provide the expected cost and performance benefits. They may also help show that these innovations do not adversely affect safety and security.

3.4.2. Examples

The following are some illustrative examples of model aided systems engineering approaches for enabling plant innovation initiatives:

(a) Passive heat removal: This passive feature (requiring no active, powered equipment after initiation) is an important means for the simplification of reactor design and for cost reduction. It has a strong impact on safety, so extensive verification and justification that it is operable in all contemplated normal, incident and accident situations is warranted. The modelling and simulation of the physics involved can provide essential evidence in conditions that would be too costly, dangerous or difficult to obtain with real physical testing. It can also be performed early in the design process and can be used to explore many different design and operational options (see Ref. [40]).

(b) Mutualized operation: To verify that this very innovative operation approach indeed satisfies the HFE principles and the needs of plant safety, some designers used co-simulation involving physical plant processes, I&C, control room human–system interfaces (HSIs) and human operators (see also Section 5.1 and Ref. [16]).

(c) Thermal constraints in integrated primary circuits: In an integrated primary circuit, the steam generators are placed within the reactor vessel. This generates thermal constraints that could be detrimental to the vessel in the long term. Modelling and simulation have been used to assess these constraints and assist vessel design to ensure they will have acceptable effects.

3.4.3. Maturity

Three dimensional finite elements modelling and simulation with safety qualified software codes is a widespread and established practice in the nuclear industry for the detailed study of reactor core neutronic and thermodynamic phenomena in normal and accident scenarios. For instance, 0-D–1-D system level modelling (e.g. with languages such as Modelica [41]) is more recent and enables multiphysics simulation of dynamic phenomena in complete plant systems or circuits. Though this practice is not as widespread in the nuclear industry as finite elements modelling, it is widely applied in other industries.

3.5. MODELLING AND SIMULATION BASED TRAINING

Modelling and simulation may be used for the training of operators in realistic conditions reflecting the dynamics of the system under their care, so that they have a better understanding of how it operates, its current state (normal or abnormal) and how to handle the system in that state. This form of training does not substitute for but rather complements other, more traditional forms of training.

The operators concerned are not only control room operators but also field operators and, more generally, operators performing missions important to the operation of a nuclear power plant (e.g. outage personnel, fire fighters, security and computer security personnel, radioprotection personnel).

3.5.1. Benefits to plant performance

Modelling and simulation based training provides many benefits to plant performance:

— Operators can be trained in realistic conditions to face situations that would be too dangerous or too costly to reproduce with the actual system.
— They can train periodically to handle exceptional situations in order to 'boost' their initial training.
— They can rehearse rarely performed procedures just before actual implementation.
— They can train in realistic conditions without affecting the availability and the operation of the actual system and of the nuclear power plant.
— They are better prepared and are less likely to make mistakes.

3.5.2. Examples

The following are some illustrative examples of model aided systems engineering approaches applied to modelling and simulation based training:

(a) Control room simulators: It is now a common practice for control room operators to train in realistic control room mock-ups associated with full scope process simulators. Such simulators may also be used to reassess operational procedures (e.g. after a modernization). In some cases (in particular in the case of plants already in operation), the full scope simulator is developed a posteriori, after the plant is fully designed or even constructed. In others, it is developed based on simulators used for plant process design.
(b) Mounting/dismounting of plant components: Some geometric CAD tools can generate models that can be simulated on handheld devices showing step by step how a plant component they helped design can be assembled and disassembled. Geometric CAD models and/or realistic 3-D scans (with lasers) can also be used to determine how large plant components can be moved in or out of buildings.

3.5.3. Maturity

Model and simulation based training is a very mature approach in all industrial sectors. In the nuclear industry, the development of training simulators for control room operators has become an integral part of nuclear power plant development. Training simulators for other operators are less common.

3.6. MODEL SUPPORTED OPERATION — PROCESS DATA RECONCILIATION

Process data reconciliation involves using digital process models (generally, but not necessarily, developed during design) and measurements provided by process instrumentation during operation to automatically:

— Identify failed or miscalibrated equipment, based on known physical limits and comparison of redundant or physically interdependent process measurements.
— Discard grossly incorrect process measurement data.
— Reduce measurements' uncertainty margins by taking advantage of redundant and/or correlated process measurement data (with the notion of 'virtual instrumentation').
— Estimate the most likely state of the process before deciding on the next course of action.

3.6.1. Benefits to plant performance

Process data reconciliation may be used to:

— Find 'missing' megawatts to increase electrical power output.
— Improve the effectiveness of the process.
— Facilitate diagnostics and maintenance.
— Improve outage planning.

3.6.2. Examples

The following are some illustrative examples of model aided systems engineering approaches applied to model supported operation through process data reconciliation:

(a) Process data reconciliation for power uprate: See annex VII of Ref. [42].
(b) VALI III [43]:
    (i) VALI III is a nuclear power plant process data reconciliation software tool. It was applied to boiling water reactors (BWRs) and PWRs in Germany and Switzerland. When applied to coolant temperature, it revealed that the measured temperatures are approximatively 1 K higher than the reconciled temperatures. When applied to feedwater mass flow, it revealed a deviation of 2% between measurements and reconciled measurements. Without installation of additional instrumentation, power uprates of more than 30 MW were achieved. These industrial applications showed that the industrial process can be described extremely accurately.
    (ii) The method was also applied to condition based maintenance. For one plant, the overall savings were evaluated to amount to $2 million by avoiding unnecessary calibration of feedwater flow orifices, preventing losses due to heat balance errors, reducing field instrument calibration and enabling early detection of losses or component/system degradation.

3.6.3. Maturity

Process data reconciliation is applied by some in the nuclear industry. It is more widespread in other industries.

3.7. MODEL SUPPORTED OPERATION — PROCESS DATA ASSIMILATION

Process data assimilation aims at estimating as accurately as possible the state of a process as it evolves in time by combining different sources of information, in particular digital process models and available process measurement data. The challenge is to keep the estimation 'on the tracks' in the face of sparse, imperfect measurements, by constant correction with fresh observations.

3.7.1. Benefits to plant performance

Process data assimilation may be used to:

— Facilitate and help optimize operation by a better, more accurate understanding of process state and behaviour.
— Improve the accuracy of faster-than-real-time 'what if' simulation aiming at predicting the effects of given courses of actions and at identifying an effective one to bring the process to a desired state. This in turn can help reduce the potential for errors in operation.

3.7.2. Example

One illustrative example of a model aided systems engineering approach applied to model supported operation through process data assimilation is a reactor co-simulator.

A reactor co-simulator is a real time simulator running in parallel and synchrony with the actual reactor core. Its main purpose is to estimate non-measurable or non-measured process variables of interest to advanced reactor control (and thus to improved plant performance), such as nuclide concentrations, reactivity components and spatial distributions of these values, as well as reactivity coefficients.

3.7.3. Maturity

Data assimilation is a well studied approach that was initially developed for weather forecasts. It is used by some in the nuclear industry but not as widely as it could be.

3.8. MODEL SUPPORTED OPERATION — RISK INFORMED OPERATION

Probabilistic safety assessment (PSA) models may be used to provide plant operation and management with risk informed insights on:

— Acceptable duration of safety equipment outages and of fallback modes;
— Periodic test intervals;
— Planning and scheduling of preventive maintenance.

3.8.1. Benefits for plant performance

Risk informed operation may help avoid unnecessary plant mode changes and optimize maintenance activities.

3.8.2. Examples

The following are some illustrative examples of approaches applied to model supported operation through risk informed operation.

3.8.2.1. Reliability centred maintenance

The overall aim of reliability centred maintenance is not necessarily to reduce the cost of maintenance programmes but to improve the functional performance and availability of plant components by optimizing the number and duration of their outages. The key is component and equipment maintenance at the right moment with an appropriate effort in terms of resources and costs. It proceeds by:

— Identification of components and their failure and degradation modes that have an impact on safety or plant performance and that need more comprehensive maintenance;
— Optimization of the maintenance strategy;
— Installation of suitable condition monitoring equipment.

3.8.2.2. *Living probabilistic safety assessment and risk monitors*

A living PSA (see Ref. [44]) is a continuously updated PSA reflecting the current plant design and operational features. It takes into account, for example:

— Modifications of plant systems and components;
— Modifications in operating manuals, periodic test strategies and maintenance strategies;
— Updated component reliability data and initiating event frequencies;
— New findings from process studies;
— The operational status of plant components and systems (e.g. the fact that a redundant division of a safety system is under maintenance or in a failed state).

A living PSA is used, for example, to assess the safety of contemplated changes to plant design, operation or training programmes. It also provides the basis for the application of risk monitors. A risk monitor is a plant specific, real time analysis tool that supports operational and maintenance decisions by determining the instantaneous risk, based on the actual status of systems and components. It ensures an optimized safety level of the plant and visualizes the consequences of several decision paths, providing operators with point-in-time risk status. Risk monitors are used today in some nuclear power plants to support operational decisions, such as to:

— Manage plant operational safety.
— Support scheduling activities (e.g. outage planning).
— Achieve greater flexibility in plant operation.
— Provide justifications for carrying out at-power maintenance.
— Provide information on the risk importance of equipment in or out of service.

A risk monitor can be off-line or on-line. Off-line risk monitors are used for planning maintenance outages, long term risk profiling, analysis of cumulative risk, evaluation of unplanned events and lessons learned from operating experience.

On-line risk monitors are used by operators for qualitative and quantitative risk monitoring. Whereas PSA provides the average risk of core damage frequency (CDFavg) based on average initiating event frequencies and unavailability due to maintenance, risk monitors provide point-in-time risk CDF(t), taking account the current plant configuration and environmental factors (see Fig. 4).

3.8.3. Maturity

Living PSA is a widely used methodology and is even required by some Member States (e.g. see Ref. [45]). To date, off-line risk monitors are widely used for planning future maintenance outages, evaluation of unplanned events including lessons learned feedback, long term risk profiling and analytics of cumulative risk. On-line risk monitors are under development (e.g. to provide operators with precise information on the unavailability of the components and the impact on plant safety).

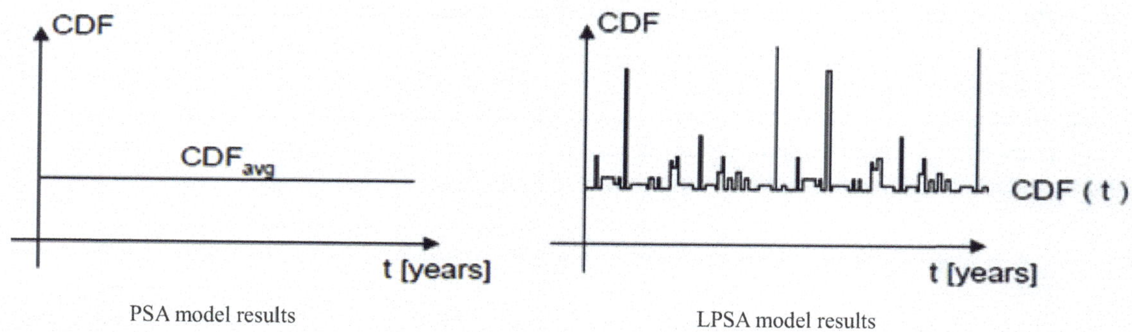

FIG. 4. Probabilistic safety assessment (PSA) model results (left) compared with living probabilistic safety assessment (LPSA) model results (right), as calculated by a risk monitor (courtesy of T. Salnikova, Framatome, Germany).

3.9. MODEL SUPPORTED OPERATION — FLEET LEVEL ASSET MANAGEMENT

Utilities operating multiple nuclear power plants (some of them possibly composed of multiple units) connected to the same power grid may consider model assisted fleet management — for example, for deciding:

— When each unit should be in outage and for what type of outage;
— The sharing out of power production between the units if they are not in baseload production;
— How many spares of each type should be provisioned and where they should be placed geographically so that they can be provided in time to the units that need them.

Models of different natures may need to be considered and integrated, such as:

— Power demand models;
— Economic models;
— Probabilistic models;
— Human resource management models;
— Environmental condition models (e.g. determining cold source conditions).

Fleet management is addressed in more detail in Section 5.2.

3.10. MODEL SUPPORTED OPERATION — MODEL BASED INSTRUMENTATION AND CONTROL

Modern I&C can embed simplified process models to better estimate the real state of the plant and of its systems and to estimate non-measurable process variables.

3.10.1. Benefits to plant performance

Model based I&C can contribute to plant flexibility and to normal operation with:

— Operation with smaller margins, closer to process optimum;
— Reduced burden on operators due to improved automation;

— Higher core loading flexibility;
— Reduction of risk of fuel damage also due to improved automation.

3.10.2. Examples

The following are some illustrative examples of approaches applied to model supported operation through model based I&C:

(a) Calculation of safety limits by I&C: Some safety limits, such as those concerning PCI, have been modelled within the I&C of several PWRs. The calculated values can not only be used for limitation functions but can also be visualized in a larger context. As part of the axial power distribution profile from in-core instrumentation, the calculated values provide operators with an excellent instrument for all non-baseload regimes (see also Annex III).

(b) On-line xenon calculation and prediction: Xenon effects impact core reactivity and plant controllability. Models based on the decay chains of iodine and xenon isotopes and a simplified reactor core (to allow implementation in safety I&C) were used in some operating nuclear power plants to improve the assessment of these effects. Such models were also used in dedicated I&C functions (e.g. rod position control or axial offset control), allowing the functions to take account of xenon effects. Because upcoming xenon effects can also be estimated, control rod movements and usage of boron and demineralized water could be optimized, reducing the cost of flexible operation and the volume of effluents (also improving environmental performance) (see also Annex III).

(c) On-line boron calculation for safety injection: In several operating PWRs, a model for boron or demineralized water injection was integrated in the I&C to compensate for the dead time of the injection system so that an underdrive or overdrive is avoided (see also Annex III).

(d) Adapted reactivity control coefficients: In some operating PWRs, at the beginning of a cycle, coefficients for reactivity control (e.g. coolant reactivity, power reactivity, control rods reactivity) of the new core are determined based on burnup dependent 3-D neutronics calculations, taking account of core history. These adapted coefficients improve control accuracy (in terms of time and amplitude; see Annex III), which in turn may be used to increase core loading flexibility and, ultimately, core utilization. They also help to minimize the number of control rod steps (reducing wear of the control rod drive mechanisms) and to optimize boron usage (reducing costs and improving environmental performance).

(e) Turbine stress evaluator: A turbine stress evaluator calculates in real time the mechanical stress of the turbine shaft that is mainly due to the temperature difference between the shaft axis and its surface. The surface temperature can be easily measured, but it is difficult to measure the temperature within the shaft. Instead, a turbine stress evaluator simulates the heat exchange within the shaft in real time and in this way estimates the temperature of the shaft axis.

3.10.3. Maturity

Model based I&C is becoming increasingly common in modern nuclear power plant I&C systems.

3.11. MODEL SUPPORTED OPERATION — IMPROVED MARGIN MANAGEMENT

Improvements in the reliability of components, identification and elimination of weaknesses in system design, availability of operating experience data and application of more sophisticated analytical capabilities enable a more detailed view on effective safety margins. This could result in margin recovery that power plants might benefit from. In practice, a safety limit that leads, in general, to a reactor trip when challenged is protected by an operational limit defined by a maximum deviation from the actual operating point that is acceptable during normal operation. This distance is called the operational margin and directly reflects the ability of operators to react to load follow demand on short notice or to other unforeseen events.

3.11.1. Benefits to plant performance

Plant operators may use the increased margins in different ways, such as:

— Maintaining the current operational limit and using the greater operational margin to operate in flexible mode;
— Keeping the same operational margin and benefiting from an improved operational limit to support more efficient fuel cycle strategies;
— Supporting power uprate projects.

3.11.2. Examples

The following are some illustrative examples of approaches applied to model supported operation through improved margin management.

3.11.2.1. Best estimate accident analyses, considering the fuel rod bow

In the analysis of a large break loss of coolant accident (LOCA) with non-nominal water gaps due to the fuel bow effect, the deterministic methods consider a conservative high water gap between adjacent assemblies, which results in significantly reduced margins to the acceptance criteria. The application of a best estimate plus uncertainty (BEPU) method allowed the recovery of margins.

In Germany, new regulatory requirements mandated BEPU analyses covering the full set of fuel rods, showing that no more than one fuel rod would exceed the acceptance criterion, making standard best estimate large break LOCA methodologies less feasible.

A full core Monte Carlo method was developed, relying on a preselection procedure to reduce computational effort from more than 400 000 Monte Carlo runs to 200 [46]. The influence of non-nominal water gaps was quantified compared to a nominal case, based on detailed full core measurement results, consideration of non-nominal water gaps in relevant tools and the application of statistical methods [47]. Such a tool could also be used in the future to identify the impact of other specific phenomena if necessary.

3.11.2.2. Power uprate

The following are two examples from PWRs in Germany where application of advanced methodologies strongly contributed to the success of power uprate projects by identifying existing margins:

— Core thermohydraulics: The impact on core thermohydraulics is often the limiting factor for a power uprate. One criterion is the departure from nucleate boiling in the case of a locked primary circuit pump rotor. Conservative assumptions assume immediate full stop of the mass flow in the concerned loop, but advanced 3-D modelling of the reactor core showed the absence of departure from nucleate boiling in such a case after the power uprate.
— Primary system structural integrity: A further limiting issue was the resulting stress in the pressure release vessel of the primary circuit in the case of an anticipated transient without scram event. Such an extended design event led to inacceptable primary stress level when applying a conventional analysis method. An advanced finite element method for most impacted locations (e.g. access ports) made it possible to perform the required justification.

3.11.3. Maturity

Advanced methodologies, such as BEPU, are becoming widely applied in safety analyses. The BEPU method explicitly evaluates uncertainties and gives insight into the considered scenarios, often leading to increased margins for acceptance criteria.

3.12. ADVANCED RADIOLOGICAL SERVICES

Advanced radiological services, which may be provided locally and/or from a remote support centre, concentrate and process the operation and radiation data of a nuclear power plant or fleet of nuclear power plants using mathematical models, enabling predictions on radiation with minimized uncertainties. Recent developments allow real time integration of plant operation and meteorological data.

3.12.1. Benefits to plant performance

Advanced radiological services:

— Contribute to workers' safety during day-to-day operation;
— Contribute to environmental performance with reliable and precise predictions in case of an accident or emergency situation;
— May be used in the training for accidents and emergencies;
— May automatically generate annual normal operation release reports.

3.12.2. Examples

The following are some illustrative examples of tools and systems used in the provision of advanced radiological services:

(a) Central radiological computer systems: These systems help assess the radiological consequences of airborne and water-borne activity releases in a plant's vicinity. Their purpose is to assess and to evaluate radiological conditions during plant operation and to reduce the routine workload of the staff collective dose rate. In accident or emergency situations, they support decision making on measures to be taken with time dependent estimates and predictions of source term and dispersion, as shown in Fig. 5.
(b) Autonomous radiation survey robots: Currently, radiological contamination surveys are primarily performed manually by operators in the field. With the advances in robotics, autonomous radiation survey robots are now conceivable. Functional specifications for such robots can be found in Ref. [48]. A detailed report on a demonstration of an autonomous radiation survey drone can be found in Ref. [49].

3.12.3. Maturity

More than 20 central radiological computer systems have been deployed, in new builds and in operating plants (see also Ref. [50]).

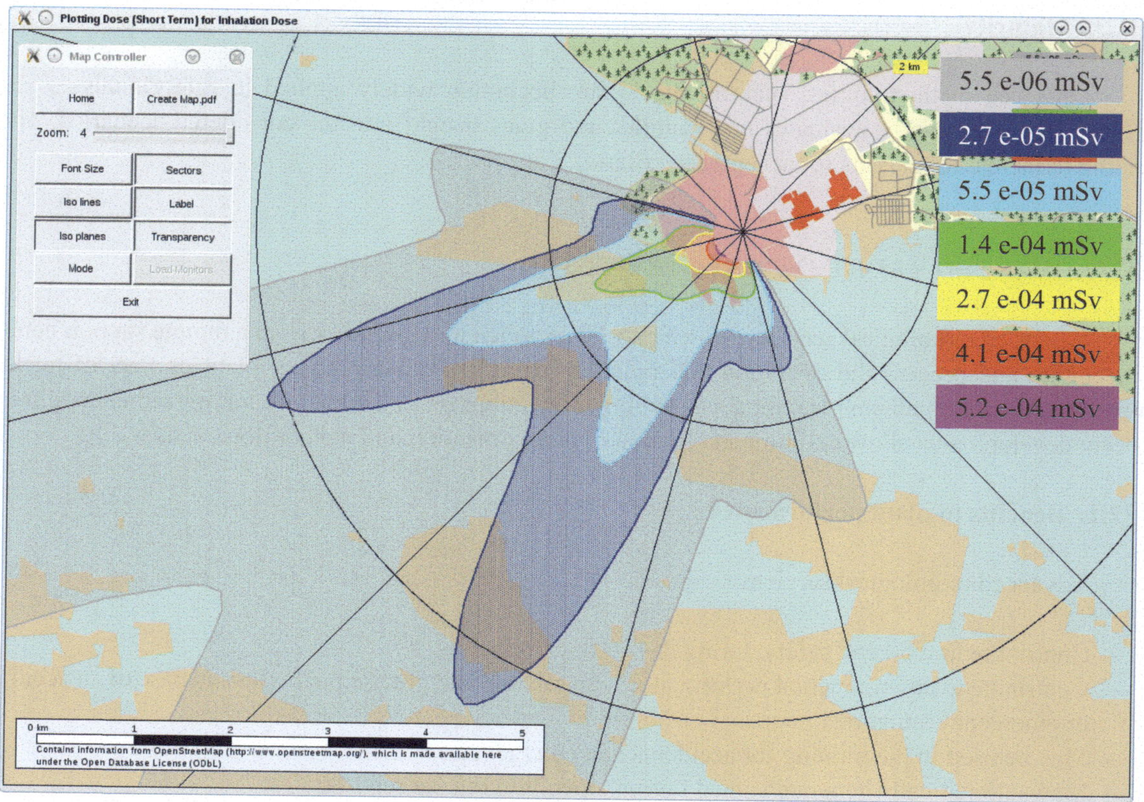

FIG. 5. Central radiological computer system dose rate dispersion display map with colour codification (courtesy of F. Weser, Framatome, Germany).

3.13. KNOWLEDGE MANAGEMENT ALONG THE PLANT LIFE CYCLE

Nuclear power plants have long life cycles; with conceptual studies, design, construction, operations and decommissioning, they could reach 60–80 years. This far exceeds the professional career of any individual. Thus, maintenance of the knowledge necessary to operate, maintain, renovate and finally deconstruct a plant is a key issue, in particular considering the importance of safety and security.

The following types of knowledge need to be considered, each with tailored strategies needed for their maintenance and long term accessibility:

— Disciplinary knowledge is not plant or product specific; it is attached to a particular trade or engineering discipline.
— Technological knowledge is also not plant specific; it is attached to a particular technology or product.
— Operational knowledge is generally plant or product specific; it is concerned with such things as how to do, how to recognize, how to diagnose.
— Engineering, safety and security knowledge about a plant and its systems is plant specific.

Basic disciplinary and technological knowledge is acquired in educational institutions or through professional training. Operational knowledge is generally acquired through professional training. With practice and experience, basic disciplinary, technological and operational knowledge may become expertise. Basic engineering, safety and security information can be found in a multitude of publications and large engineering databases. However, knowledge is obtained only by fully understanding the vast amount of information available.

3.13.1. Benefits to plant performance

Certain types of models may help to translate information into knowledge, since by definition they are a selective and simplified representation of an actual system, entity, phenomenon or process. Their simulation may also help, as it can show how a system, entity, phenomenon or process evolves in time, depending on initial and boundary conditions.

Naturally, this may be so provided that the models are kept up to date, that the associated tools are themselves maintained and kept operational and that the concerned personnel are adequately trained.

3.13.2. Examples

The following are some illustrative examples of tools and models supporting knowledge management along the plant life cycle:

(a) I&C functional diagrams: I&C functions are often expressed in functional diagrams, which are a form of digital models that can be simulated and then can be used for software code generation. Simulation allows I&C engineers to better understand what is specified and can reveal unexpected and sometimes undesirable behaviours and/or features; such verification is an integral part of the I&C development process and is called functional validation.
(b) System level physical and process models: Increasingly, such models are developed to understand the behaviour, effects and interactions of complete plant systems in graphical languages, such as Modelica [41] or Simulink. With their simulation, future generations will also be better able to understand the functioning and operation of the associated plant systems.

3.13.3. Maturity

Modelling and simulation are widely used in the nuclear industry, though their application could be more systematic and cover more activities. However, they are still rarely mentioned in plans for the long term management of plant specific engineering, safety and security knowledge.

3.14. EFFICIENCY OF INDIVIDUAL TEAMS

In traditional engineering approaches, concepts, systems, system states, requirements, solutions and boundary conditions are described in 'static' documents using natural language and informal diagrams. Although this is generally a good way to convey overall intentions, the lack of formality and rigour often leads to ambiguity, incompleteness and errors. Also, in the case of large and complex systems where many different configurations and situations need to be taken into account and where undesired effects can propagate far from initial causes or through subtle mechanisms, fully understanding and deriving all the consequences of what is expressed is not easy.

In model aided engineering, traditional documentation is complemented with models that bring structure and abstraction for easier understanding, precision and rigour for reduced ambiguity and errors. There are different types of models, for example:

— Geometric CAD is a technique to create digital 2-D and 3-D models defining the shape, material, assemblage and kinematics of physical objects and structures. This is usually much faster, more efficient and less costly than constructing and working with physical mock-ups.
— I&C functional diagrams are models specifying required I&C functions in terms of predefined elementary functions. They offer a high level graphic representation that is easier to understand than software code written in general purpose programming languages (e.g. the C language). Also, simulation allows I&C engineers to verify right from the early stages of the I&C engineering process

that the specified I&C functions are indeed what they have in mind. Lastly, translation of functional diagrams into executable code can be much assisted by code generation tools.

3.14.1. Benefits to plant performance

Models can also benefit from extensive support by software tools, in particular but not only, for simulation, analysis and information retrieval, which can greatly improve the efficiency of individual teams and engineering disciplines, for example by:

— Facilitating the understanding of complex systems. Simulation can 'animate' a behavioural model and show its precise meaning and effects in various situations. Navigation and query tools can facilitate the retrieval of information in models more effectively and accurately than in natural language and informal diagrams. This may further reduce the number of engineering errors and shorten engineering schedules.
— Revealing errors early in the engineering process. Experience in all industrial sectors shows that the later a fault is revealed, the higher the cost it incurs and the longer the time necessary to correct it. In the worst cases, faults are revealed during operation, possibly with very negative or even catastrophic effects. Simulation and analysis can reveal errors early in the development of solutions, long before they are implemented.
— Providing an efficient, rigorous framework and basis for the verification and validation of implemented solutions.

3.14.2. Example

One illustrative example of a model aided systems engineering approach that improves the efficiency of design teams is continuous design verification of a reactor component's cooling water system.

Due to the many reactor components such as system needs to cool, and to the stringent safety, reliability, fault tolerance and maintainability requirements, a reactor component's cooling water system is composed of many elements (e.g. pumps, pipes, valves, heat exchangers) that can be organized into many different configurations. The system also needs to take into account numerous normal and abnormal reactor states (where sets of components to be cooled and amounts of generated heat are different) and operator actions (e.g. for maintenance or testing). Also, because the design of a nuclear power plant is a long and iterative process, the number and needs of the 'clients' of the cooling water system may evolve, but its design cannot wait until all is stabilized. Because the system is of the highest importance to safety, when a change occurs, the design has to be thoroughly verified.

This is difficult, time consuming, costly and error prone when done manually, but requirements and system level physical process modelling and simulation were used to verify the design automatically, rapidly and effortlessly at each change, taking into account all the client's needs, all system configurations, all reactor states and all operator actions.

3.14.3. Maturity

Geometric CAD and I&C functional diagrams have been applied to improve engineering efficiency for many decades and throughout all industrial sectors, including the nuclear industry. Other forms of efficiency-improving modelling (e.g. system level physical process modelling with languages such as Modelica [41]) are also widely applied in non-nuclear sectors, but less widely and less systematically in the nuclear industry.

3.15. EFFICIENCY OF MULTIDISCIPLINARY AND COLLABORATIVE ENGINEERING

The essence of systems engineering (see Refs [22, 35, 36]) is the art of coordinating (as necessary and only as necessary) multiple individuals, teams, engineering disciplines, organizations and stakeholders. Because these stakeholders may have very different cultures, viewpoints and approaches to problems, and they may be distributed over vast distances and time zones, they often have difficulties understanding one another. Model aided systems engineering uses models as a means for rigorous, accurate and timely information exchange and coordination of all concerned parties along the life cycle of a system (e.g. the nuclear power plant, a plant system or the I&C system). Indeed, even though documents based on natural language and informal drawings are useful in the very first and conceptual stages of engineering, they are generally not precise enough to ensure a streamlined coordination in the engineering of a system as complex as a nuclear power plant. Extensive tool support that can help bridge the differences in culture and approaches is needed.

Like for all artefacts developed and used in long and complex engineering processes involving numerous parties, a technical infrastructure (e.g. adequately interfaced information systems and configuration management systems) is also needed to ensure the storage, protection, retrieval and communication of models.

3.15.1. Benefits to plant performance

Experience shows that, for large and complex systems, efficient work in silos, as discussed in the preceding section, is not enough; inadequate coordination of concerned parties generally leads to inefficient systems engineering, late error discovery and correction, delays, cost overruns, less-than-optimal system performance and, in the worst case, accidents. Evolutionary approaches (whereby a new design or solution is obtained by introducing limited changes to a proven one) has for a long time allowed the nuclear industry to do without or with only very limited model supported coordination but, as other sources of energy are becoming ever more competitive, it now needs to innovate more radically while still ensuring extremely high safety and security.

Models may be used to define interfaces and interactions between entities under the responsibility of different parties precisely and unambiguously. Such models are sometimes called 'contracts' and the approach 'contract based systems engineering'. Once a contract is agreed by all concerned parties, each party can proceed separately until they need to reconvene to refine the contract as engineering progresses or to amend it if some parties meet difficulties. This facilitates interdisciplinary work, and as tool supported mechanisms ensure better consistency and completeness checks, changes can be implemented more quickly and controlled and communicated more effectively.

3.15.2. Examples

The following are some illustrative examples of approaches that enhance the efficiency of multidisciplinary and collaborative engineering.

3.15.2.1. Geometric computer aided design/manufacturing

Geometric CAD may be used at early engineering stages to specify the geometric interfaces between objects and structures that need to be assembled and also, when necessary, the geometric envelopes within that each need to fit in.

Computer aided manufacturing (CAM) is a technique exploiting geometric CAD models to develop efficient and accurate production (manufacturing or construction) and verification processes. Geometric CAD and CAM are nowadays supported by extremely powerful and extensive software tools helping ensure, right from the early engineering stages, that objects and structures can be produced, verified individually and assembled even when geometric design is distributed among many teams and organizations.

Advanced geometric CAD/CAM software tools can produce animated sequences showing how to assemble and disassemble composite objects and how to mount and dismount objects on their supporting structures, thus preparing the way to installation at the site and maintenance.

3.15.2.2. Instrumentation and control engineering with an evolving engineering simulator

I&C systems and operator displays are often validated by co-simulating I&C models and plant process models. As late error detection often leads to unplanned costs and missed deadlines, validation is best performed as early as possible. Unfortunately, full scope simulators and operator training simulators are usually available only very late in a project. Light and evolving full scope simulators (i.e. simulators kept up to date as engineering decisions are made) accessible right from the design phases of a project can be used to support early and iterative validation and error detection. Reference [51] illustrates the steps of the approach that is becoming a standard solution for both new builds and operating plants worldwide.

3.15.2.3. Coordinated engineering for a reactor component cooling water system

To continue with the example of Section 3.14.2, the interface between the cooling water system and each of its 'clients' was specified by a contract specifying the temperature range within which the client has to be maintained, the maximum heat produced by the client, the specific heat of the client and the constraints set by the client's heat exchanger (some of them depend on reactor state). Once all these contracts were agreed upon, the teams in charge of each system could confidently work on their own. From time to time, as the engineering of clients progressed, some contracts had to be amended and/or extended but then it was always clear with whom to renegotiate.

Once the process design team for the cooling system had a detailed design model for it, identifying all of its components and their interactions, a dependability team was able to extend the model, introducing the failure modes of each component. With this extended model, they were able to perform probabilistic analyses estimating the failure probabilities of the system. Using simulation, they were also able to automate the mandatory FMECA, a notoriously time consuming and costly activity. Each time the detailed design was modified, probabilistic analyses and FMECA could be repeated with minimal effort and cost.

3.15.2.4. Design optimization

Design optimization refines the design of a component in a series of engineering steps. Each step adds new information as data are collected and engineering decisions are made. A possible approach is to start addressing physical aspects in concurrent experimental and numerical simulation tracks as shown in Fig. 6 [52], prior to using physical prototypes. This helps limit upfront investment in costly physical prototypes and unworkable designs. Concurrent exploration of the two tracks also allows them to mutually support one another, with exchange of information and results in both directions.

3.15.2.5. Preparation of procurement and construction

A 3-D design model of the Hualong One nuclear power plant features more than 50 000 pieces of equipment, 165 km of piping, 2200 km of cables, etc. With such amounts and due to frequent modifications, manual procurement planning is a daunting and error prone task. With the 3-D model as a tool friendly and always up-to-date reference, this can be done much more efficiently and with much fewer errors.

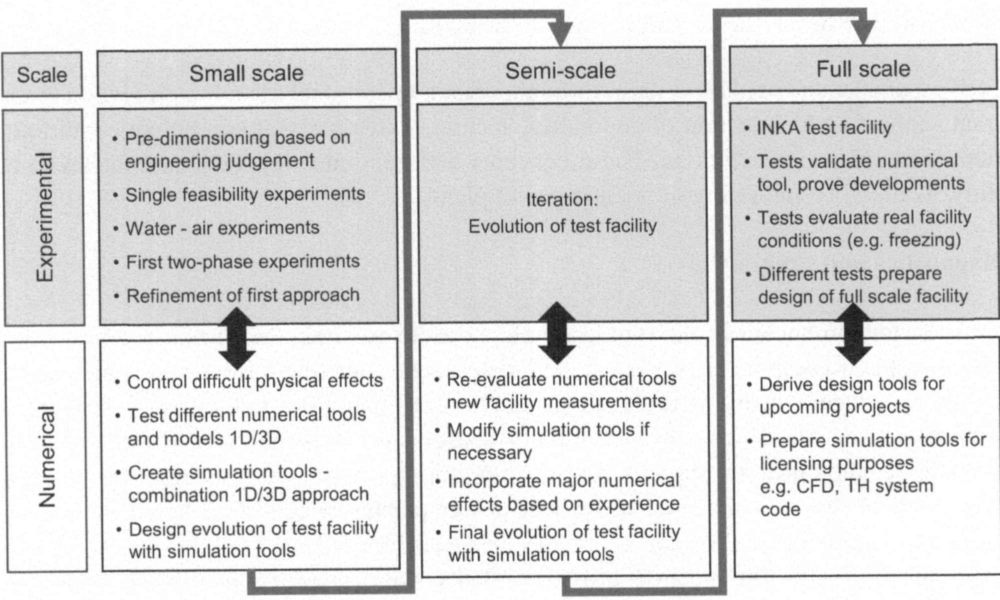

FIG. 6. Design process for component design optimization (adapted from Ref. [52]). CFD — computational fluid dynamics; INKA — Integral Test Facility Karlstein; TH — thermohydraulic.

The 3-D model also provides on-site construction teams with up-to-date design data for piping and cabling so that, with the help of scheduling and schedule simulation, they can prepare and coordinate their interventions.

3.15.3. Maturity

All industrial sectors use models for improved multidisciplinary and collaborative engineering. However, in the nuclear industry, work in silos (with insufficient coordination and interfacing between engineering disciplines and organization) is still widespread; there is thus a large potential for improvement.

4. IMPROVED DIAGNOSTICS AND PROGNOSTICS

Although diagnostics solutions have been around for decades, the recent maturity of digital technologies and the wide availability of plant data enable a paradigm shift in condition monitoring that moves beyond the conventional systems engineering framework and transforms nuclear utilities into data driven organizations. Diagnostics and prognostics provide operators with comprehensive, context rich and accessible insights to support condition based maintenance and enhanced asset management, thus improving plant performance. Diagnostic and prognostic solutions are simultaneously leveraging plants' data, operator expertise and advanced technologies, such as signal processing techniques, modelling and simulation or AI techniques. They are often referred to as 'digital twins' for their ability to mirror plants' complex designs and behaviours.

4.1. MAIN NOTIONS

Effective management of plant systems throughout their operational life relies on several fundamental concepts that support early detection of anomalies, accurate assessment of equipment condition and the optimization of maintenance activities. These concepts underpin contemporary approaches to ensuring the reliability, availability and safety of nuclear power plants.

4.1.1. Diagnostics and prognostics

Diagnostics and prognostics consist of using plant data to ascertain the current and future condition, respectively, of a plant component or system. The data needed for diagnostics and prognostics can be provided both by process instruments, such as neutron, temperature and pressure sensors, and by diagnostic instruments, such as acoustic and vibration sensors (see Refs [53, 54]).

Diagnostic technologies to support on-line monitoring, noise analysis and vibration analysis have been around for decades and used in nuclear power plants for condition based maintenance and troubleshooting of equipment and system anomalies (see Refs [55, 56]).

Combining this with prognostics provides plants with ample capability to optimize their maintenance schedules, avoid unnecessary work and replace ageing components based on diagnostic and prognostic data rather than number of years of service. In addition, if qualified, the built-in diagnostics of digital I&C platforms and systems could be used to replace or reduce some manual surveillance testing (see Ref. [57]). The use of qualified diagnostic I&C functions could also reduce the risk associated with manual surveillance testing. Importantly, implementation of diagnostics and prognostics can help plants reduce outage time, avoid plant trips and improve efficiency. The methodologies and technologies used to perform diagnostics and prognostics of a system can range from manual analysis by a subject matter expert to the most advanced AI technologies.

4.1.2. Open minded condition monitoring

Condition monitoring consists of the observation of an installation and its systems, components and structures, and of the analysis of the information collected to detect signs of abnormal behaviours or conditions. It also includes monitoring processes at the heart of the nuclear power plant. Early anomaly detection and mitigation are essential before consequences are too detrimental. However, though condition monitoring is traditionally aimed at detecting harbingers of component failures, as can be seen in Refs [55, 56, 58–60] and as discussed in this section, it is important to consider that initially it is not known whether the anomaly is due to structure degradations, component failures, process instrumentation (measurement) error, extreme beyond-design events, human error or malicious attacks. Thus, to correctly identify the origins of the deviation, an analysis needs to be done from multiple perspectives.

For example, in a digital I&C control system, a data communication storm was observed in one data communication link. That could have been an effect of a successful computer security attack, but analysis showed that it was caused by an electronic component failure.

4.1.3. Condition based maintenance

Condition based maintenance consists of assessing the need for equipment maintenance and scheduling maintenance activities based on the equipment's operability rather than at predefined time intervals. The condition would be assessed based on the insights gained from performing diagnostics and prognostics. In the specific case of maintenance activities anticipating future degradation of component operability, 'predictive maintenance' is the preferred term.

Condition based maintenance (see Ref. [59]) is an optimized variant of preventive maintenance. It is not performed at regular intervals but only when certain criteria (which depend on the type of equipment) indicative of degradation and upcoming failure are reached. It is not always possible; one main difficulty is the identification of criteria specific enough to give an unambiguous and accurate indication; another is that these criteria need to be monitored on-line, but that is not always the case.

The main advantage of condition based maintenance, when it is possible, is a better optimization of maintenance:

— Equipment can be used to its full potential while avoiding failure.
— Testing and maintenance activities can be reduced without risking higher failure rates.
— Maintenance activities can be anticipated and plant productivity preserved.

Also, some of the drawbacks of periodic maintenance can be avoided or reduced, as interventions on perfectly operating equipment can:

— Reduce their availability.
— Induce degradation (e.g. some tests are destructive or damaging).
— Increase failure rates (e.g. due to human error).

Condition based maintenance in the nuclear industry is not as widespread as in other industrial sectors, such as the aviation industry or the petrochemical industry. Also, the existence and maturity of maintenance-triggering criteria strongly depend on the type of equipment considered.

4.2. ANOMALY DETECTION IN PLANT DATA

Anomaly detection involves the trending of station data to identify deviations from baseline behaviour. Typically, additional context is needed from plant records or forensic examinations alongside interpretation by qualified personnel to act on any detected anomalies. Anomaly detection can be used to alert station personnel of degraded operation or impending failure of plant equipment.

Several approaches can be pursued to detect anomalies at stations. In the simplest of forms, system responsible engineers and signal analysts can visually inspect raw or processed data to identify anomalous behaviour based on experience. Anomalies can also be detected automatically, either with built-in advanced pattern recognition features of commercially available plant data historians, or with dedicated software utilizing machine learning based methodologies. The general categories of methods used in anomaly detection are identified and briefly described in Table 1.

Regardless of the approach pursued, anomaly detection techniques are broadly applicable to plant process sensor signals (e.g. pressure, level, flow and temperature transmitters), diagnostic signals (e.g. accelerometers, acoustic emission sensors) and non-time-series data (e.g. chemical analysis results), provided that there are sufficient data available to establish baseline behaviour. A known limitation of anomaly detection is that any identified deviations typically need further operational context and specialist interpretation before they are actionable by station personnel.

4.2.1. Benefits to plant performance

Detecting deviations from expected equipment behaviour enables station personnel to react quickly to potential issues, reducing the frequency of unplanned outages. In addition, the implementation of anomaly detection methodologies in continuous on-line monitoring tools automates equipment health monitoring, which can reduce analyst and component specialist troubleshooting efforts. Systematic tracking of anomalies by an automated system can also facilitate reporting and trending of equipment health.

TABLE 1. APPLICATION/USE CASES BASED ON ROBOTIC PROCESS AUTOMATION AND AI

Method	Description	Advantages	Disadvantages
Threshold based anomaly detection	Acceptable ranges for parameters of interest are set based on operator experience, expert insight or statistical characteristics of the data (e.g. interquartile range); data falling outside the acceptable range are detected/identified	• Easy to implement • Explainability of detection process	• Typically unable to identify cases where equipment is trending towards a failure but still within the acceptable parameter range
General rule based anomaly detection	A generalization of the threshold based approach to include the use of complex conditional statements relating the state of multiple parameters or the use of frequency domain criteria (e.g. detecting new peaks in a frequency spectrum)	• Explainability of detection process	• Requires extensive empirical or physical a priori knowledge of the expected system/ component behaviour • The whole set of rules may need to be reformulated to account for new data that may not fit the anticipated behaviour
Machine learning based anomaly detection	The use of advanced analytics techniques, such as autoencoders or single class support vector machines, to automatically infer the normal behaviour/dynamics of the data and identify anomalous states; a more detailed description of the methods available is provided in Section 6.13	• Can infer complex dynamic behaviour without specialist knowledge • The approach is readily adaptable to new data	• Methods depend on accurate tuning of hyperparameters, which can be challenging a priori; this disadvantage can be mitigated through the use of ensemble methods • Limited explainability of detection process

4.2.2. Examples

The following are some illustrative examples of applications of anomaly detection in plant data:

(a) Rosenergoatom turbine generator predictive analytics: A predictive analytics project aimed at detecting anomalies of the turbine generator using advanced pattern recognition has been implemented at Unit 6 of the Novovoronezh nuclear power plant.

(b) Verification of calibration status of nuclear plant pressure transmitters: It is reported that more than 90% of nuclear grade pressure, level and flow transmitters maintain their calibration longer than a typical fuel cycle, resulting in unnecessary calibrations during unit outages. Eliminating unnecessary calibrations reduces the risk of radiation exposure to workers and the potential for human error, plant trips or spurious equipment actuations. In 2021, the US Nuclear Regulatory Commission approved the on-line monitoring technology to extend the calibration intervals of nuclear plant pressure transmitters by as much as 12 cycles or over 20 years in some cases (see Ref. [61]). Annex IV describes the successful use of threshold based anomaly detection to identify faulty or drifting transmitters at Sizewell B in the United Kingdom.

(c) Anomaly detection using vibration measuring sensors: See Annex II.

(d) Anomaly detection using neutron noise analysis: Using noise analysis techniques, plants have successfully detected excessive vibrations of reactor internals, identified blockages in pressure sensing lines and discovered core flow anomalies (see Ref. [56]).

4.2.3. Maturity

Anomaly detection methods are well established within the nuclear power industry, in particular for cases using statistical (e.g. interquartile range) and advanced pattern recognition.

4.3. REAL TIME PREDICTION AND ESTIMATION OF PLANT PARAMETERS OF INTEREST

Station personnel often need knowledge of parameters that are challenging to measure directly, to inform plant operations and maintenance. In several instances, surrogate parameters are already being measured that could be analysed to infer the desired parameters, based on data driven approaches or a priori knowledge of the governing physics. The approach can be used to estimate physical/process variables that cannot be measured with existing plant instrumentation (e.g. valve leakage rate) or those that are infrequently measured (e.g. neutron flux detector prompt fraction estimation, valve stem thrust/torque). It can also be used for uncertainty reduction of process measurements, by supporting process data reconciliation efforts.

Several techniques are available to estimate such parameters of interest. Methodologies can utilize analytical techniques, classical signal processing, machine learning based continuous regression, or a combination thereof.

4.3.1. Benefits to plant performance

Several parameters that indicate plant performance or component health, such as leakage rate for cycle isolation monitoring or valve stem thrust, are infrequently measured or not measured at all. Knowledge of such parameters would enable stations to assess the severity of an issue and determine if (and how fast) a component is trending towards a degraded state through on-line monitoring. Furthermore, reduced uncertainty in safety margins will enable more efficient plant operation by a reduction in the conservatism needed.

4.3.2. Examples

The following are some illustrative examples of real time prediction and estimation of plant parameters of interest.

4.3.2.1. *Heat exchangers*

Reference [62] describes an I&C device enabling the condition based maintenance of large heat exchangers. The device acquires four temperatures and two flow rates associated with the heat exchanger. With these measurements, the heat transfer coefficient of the heat exchanger can be calculated in real time. The decrease of this coefficient indicates the need for maintenance.

4.3.2.2. *Motor operated valves*

Utilities are typically required to continually demonstrate operability of their safety related motor operated valves, in particular by verifying that valve actuators produce enough thrust to actuate the valve stem. Stations typically assess the thrust margin available through at-the-valve testing performed during outages.

Alternative means are available for estimating valve stem thrust, which rely on motor operated valve currents and voltages measured at the motor control centre. Motor control centre diagnostics are available in several product offerings of nuclear suppliers and are well established in the technical community [63]. Actuator torque and valve stem thrust estimates through motor control centre diagnostics can be produced while the station is on-line, allowing station personnel to trend the available actuator margin on an ongoing basis.

4.3.2.3. Prompt fraction estimation of an in-core flux detector

For some reactor designs, licensees are required to demonstrate adequate dynamic response of their reactor instrumentation. Methods have been developed to assess in-core flux detectors' response on-line based on steady state noise data [64].

4.3.2.4. Estimation of steam generator clogging

Deposits of corrosion products in the secondary side of steam generators may result in the clogging of flow holes in tube support plates. Beyond the loss of efficiency, consequences can be critical: excessive vibration of the steam generator tubes and risk of cracks. Treatment of clogging (chemical cleaning or steam generator replacement) is very costly.

As some of its nuclear power plants were affected by the phenomenon, the French multinational energy provider Électricité de France (EDF) developed a Modelica model (see Ref. [41]) to simulate steam generator behaviour during power transients. Clogging could then be estimated by comparing measured response curves with simulated ones for different clogging levels.

4.3.3. Maturity

Methodologies for parameter estimation are widely established, and commercial offerings are available (e.g. motor control centre diagnostics).

4.4. REAL TIME FAULT IDENTIFICATION AND CLASSIFICATION

Fault identification and classification technologies are unlocking a new class of smart automatic diagnostic tools with the provision of comprehensive plant state estimators supported by context rich insights directly actionable by non-specialists (Fig. 7).

They are the cornerstone of the transformation of utilities into data driven organizations where diagnostics are performed in a single and shareable source of trust to monitor asset health and perform condition based maintenance.

Often referred to as 'expert systems' in common AI classifications or simply 'digital twins' for their ability to mirror plant behaviour, these technologies overcome the limits of data accessibility and completion by leveraging operating experience, design constraints, simulations and experts' knowledge. Fault identification and classification techniques remain highly active research areas in AI applied to

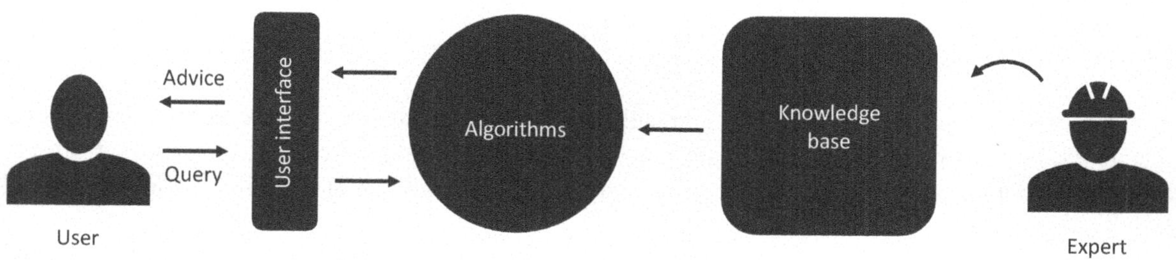

FIG. 7. Illustration of fault identification and classification frameworks (courtesy of A. Schwartz, Metroscope, France).

nuclear plants, with applications of interest related to pump seal monitoring [65] and motor control centre diagnostics of motor operated valves [66].

4.4.1. Benefits to plant performance

Fault identification and classification technologies allow for responsive, targeted and financially efficient maintenance decisions while optimizing maintenance volumes. As information is actionable and shareable with non-experts, those technologies are single sources of trust throughout the organizations where they are adopted, addressing corporate strategic needs as well as those of system responsible engineers and operating teams. Diagnostics are executed in real time and throughout the plant's history, providing a plant performance knowledge hub that facilitates knowledge management and training of new personnel.

4.4.2. Examples

The following are some illustrative examples of real time fault identification and classification:

(a) Pump seal diagnostics: Several nuclear power plants currently identify and diagnose reactor coolant pump seal anomalies and failures based on seal-vendor-provided rule based criteria. Some detailed qualitative information is provided in Ref. [65]. There is also ongoing development of an AI based pump seal condition monitoring tool at a national nuclear laboratory [66].

(b) Diagnostics of a full scope nuclear power plant secondary cycle: Examples of real time fault identification in PWR secondary loops are provided in Annex I. The technology contains a knowledge base of up to 150 fault modes for the steam cycle of plants, including cycle isolation leaks, reheater tube ruptures, condenser fouling and more.

(c) Parameter-Free Reasoning Operator for Automated Identification and Diagnosis (PRO-AID): The PRO-AID software package performs real time monitoring and diagnostics using a form of automated reasoning based on a process model derived from the fluid system piping and instrumentation diagrams and the electrical on-line drawings. To ensure the explainability of results, the reasoning is constrained by logical trees automatically deduced from the model [67]. PRO-AID has been successfully applied on plant data sets from two US nuclear utilities.

4.4.3. Maturity

Automated fault diagnostics and classification are relatively new within the nuclear industry, with limited commercial offerings.

4.5. ESTIMATING REMAINING USEFUL LIFE OF PLANT COMPONENTS

A particular case of plant parameter estimation as described in Section 4.3 is a component's remaining useful life (RUL). Although anomaly detection (Section 4.2) can alert station personnel to deviations from nominal operation, and fault classification can provide an actionable assessment of a component's condition, neither approach provides direct insight into the anticipated time to failure. The premise of RUL is to obtain a tangible metric that can inform operators of when component replacements are needed to avoid in-service failure. Knowledge of the RUL enhances the insight gained from conventional wear and reliability studies, as it is linked to the dynamic behaviour of ageing plant components. The RUL is expected to be a key enabler of plant long term operation and predictive maintenance, as it permits time targeted utilization of maintenance resources to proactively mitigate component failures.

Several methodologies have been demonstrated and employed to derive RUL estimates for industrial components, including classical probabilistic techniques, such as Markov chains, or data driven machine learning techniques. Methodology development for RUL estimation remains a highly active research area in the machine learning and AI research communities.

4.5.1. Benefits to plant performance

Maintenance cost savings are the primary benefit of RUL estimates, as they can be used to prioritize maintenance activities and shift periodic maintenance tasks to condition based maintenance. The RUL is expected to improve plant availability by avoiding unplanned outages due to failure of the monitored component/system.

4.5.2. Examples

The following are some illustrative examples of estimating the RUL of plant components:

(a) Machine learning to predict a wind turbine bearing's RUL: A machine learning approach was applied to vibration signal data to determine the RUL for a degraded bearing on a wind turbine [68]. A Weibull hazard rate function was used for calculating the RUL of the bearing.
(b) Cable RUL: For example, prognostics have been used with reasonably good success in predicting the RUL of polymers that are used as cable insulation material in nuclear power plants. This work alone has, in one instance, saved a nuclear plant nearly $4 million in cable replacement costs (see Ref. [69]).

4.5.3. Maturity

Remaining useful life applications are relatively new within the nuclear industry, with limited commercial offerings.

5. OTHER MEANS OF IMPROVING PERFORMANCE

In addition to the model aided systems engineering concepts and the improved diagnostics and prognostics discussed in the previous sections, a range of complementary approaches can also improve nuclear power plant performance, as discussed in this section. These include strategies for mutualized operation of multiple reactors and improved fleet management, as well as decision making aids that support operators in determining the appropriate actions under varying plant conditions. Advanced technologies such as robots and drones can enhance inspection and maintenance capabilities, while fault tolerant control systems strengthen operational resilience. Remote support solutions, along with computer security monitoring and incident management systems, can contribute to safer and more reliable plant operation. On-line operator documentation and workflow optimization tools help to streamline processes and reduce human error, and dedicated measures to support long term operation help to ensure sustained performance and asset integrity across the plant life cycle.

5.1. MUTUALIZED OPERATION OF MULTIPLE REACTORS

Personnel costs represent a significant part of the operation costs of a nuclear power plant. It is particularly true for control room and field operators, who work in 24/7 shifts and who also need periodic training to face infrequent situations, including accident situations.

Recently, with innovations such as simplified plant processes; passive systems and features; aids from advanced digital technologies; condition based maintenance; and sometimes smaller, more homogeneous reactor cores (as is the case for SMRs), workloads have been reduced to a point enabling mutualized operation, whereby multiple units in a plant are operated by a single team of control room operators together with a single team of field operators, including accident situations.

5.1.1. Benefits to plant performance

Even though extensive HFE studies are necessary to justify the safety of the approach, mutualized operation reduces personnel costs significantly. It also reduces construction costs, as control rooms are also mutualized.

5.1.2. Example

One illustrative example of mutualized operation of multiple reactors is NuScale. The simplified design of the NuScale SMR allows mutualized operation of up to 12 units by a small team from one central control room. This reduces the production costs for a NuScale plant with multiple modules or reactors very significantly.

5.1.3. Maturity

Most operating nuclear power plants consist of a single unit arrangement. Dual unit arrangements are rare, and current regulations in Member States often limit the number of units that can be controlled by the same team from a single control room. However, several SMR projects are designed for more extensive mutualized operation, and some have already been approved (e.g. NuScale).

5.2. IMPROVED FLEET MANAGEMENT

The technologies and methods described in this publication can have a substantial impact on the performance of a nuclear power plant taken individually. They can also provide substantial improvement in performance across a fleet of nuclear power plants and, more generally, across a fleet of power generation assets, by integrating insights from multiple generation facilities into a single situational view that allows a fleet to be managed as a system of assets instead of multiple individual independent assets.

As more plant component and system performance data with status information become accessible in central locations remote from any single facility, asset owners are increasingly using advanced algorithms, including machine learning and AI, to combine these data into actionable fleetwide insights.

5.2.1. Benefits to plant performance

These insights are used to:

— Optimize and implement generation dispatch strategies that combine multiple market and asset status criteria.
— Establish optimized maintenance, logistics and service dispatch at a fleet level that eliminates redundant resources and advances plant performance by using condition based maintenance strategies.
— Provide a fleetwide/countrywide computer security and physical security view that can identify and mitigate multiprong adversary offensives.
— Support decision making regarding lifetime extension and upgrading of existing assets and construction and features of new assets.

5.2.2. Examples

The following are some illustrative examples of improved fleet management.

5.2.2.1. *Dynamic economic, carbon impact and risk based dispatch*

A dynamic dispatch framework integrates complex inputs across multiple nuclear and non-nuclear generation assets to allow decision makers to balance trade-offs and quantify benefits from increased flexibility and improved forecasting. The penetration of renewable generation assets to lower the global carbon footprint has created multiple, diverse and often competing criteria for asset dispatch selection. Dynamic dispatch using integrated plant performance data allows forward contracting for production capacity while considering the full range of operational uncertainties in generation, demand, forecasts, prices, carbon impact and the risks of unmet demand or excess generation in real time. This provides systems operators with the tools for reducing uncertainty, addressing increasing costs and identifying opportunities for corrective actions at future decision points as those decision points approach.

Dynamic dispatch integrates multiple uncertainties into a unified framework and accepts all kinds of probability distributions and machine learning insights. This dynamic optimization approach is comprehensive and considers the flexibility of response actions taken at later decision stages, when updated information and improved forecasts become available. A reliability constraint, based on plant performance data, is accommodated directly in terms of the power balance between supply and demand in real time.

5.2.2.2. *Fleet on-line monitoring, diagnostic and security centres*

To continue to advance nuclear power plant performance in multiple dimensions, owners are concentrating plant performance and status information into centralized, remote support centres addressing equipment performance monitoring, diagnostics, prognostics and analysis, and security functions. This allows scarce resources to be leveraged across multiple sites and allows support centre experts to provide site personnel with recommendations and anticipated warnings based on the evaluation and interpretation of site data, signal abnormalities and explicit defect patterns.

This supports high work efficiency and provides common insights that can be used across multiple assets simultaneously, enabling centralized maintenance and logistics planning. Incipient failures can be identified, and maintenance resources can be dispatched using pooled personnel and parts resources. The benefit–cost ratio for the monitoring and diagnostics centre can exceed 4:1. The benefit calculation includes heat rate savings, cost-to-repair savings and a benefit for ensured generation.

Asset health management is another element of a fleetwide monitoring and diagnostic centre. This concept builds on the component and system engineering programmes that have been prevalent throughout the nuclear power industry. An asset health management programme provides the means to assess and compare the current health status of numerous dissimilar assets across a system, unit, plant or fleet. The health status includes operational trends, material conditions, performance and efficiency, maintenance strategies and actions, and many other elements that contribute to an asset's overall health. An asset health management programme also establishes the structure and processes to integrate data and information from a variety of sources, to uniformly assess these data and information and ultimately to produce an objective comparison of the health of multiple assets.

Plant performance data are increasingly being leveraged to detect potential cybersecurity and physical security anomalies and indicators of compromise. Either integrated into a monitoring and diagnostics centre or monitored at a separate security operations centre, plant performance data can improve the security awareness and aggregated security posture of a nuclear fleet.

5.2.2.3. Fleetwide monitoring and diagnostic centre

For an example of a pilot project to implement new methodologies and quantify the benefits of a fleetwide monitoring and diagnostic centre as applied to the utility's generation fleet, see Ref. [70].

5.2.3. Maturity

Large utilities have extremely strong incentives to optimize the management of their fleets (which may include both nuclear and non-nuclear assets). The use of advanced digital technologies for that purpose is widespread and long standing.

5.3. DECISION MAKING AIDS FOR OPERATION

Human error during construction, operation, outages and possibly decommissioning is a significant cause of impaired plant performance. Clear, informative HSIs as enabled by digital technologies (see Section 6.7) may help prevent errors by providing operators with a better understanding of the current situation, but beyond that, guidance in the determination of appropriate courses of action may take many different forms:

— Computerized procedures (see Refs [71–73], for example) may be used to help control room and field operators identify and apply the appropriate procedure, given the current state of the plant or plant system.
— In cases where no predefined procedure applies, models, data reconciliation, data assimilation and faster-than-real-time simulation may help control room operators determine the outcome of a given course of action, or which course of action can bring the plant or plant system to the desired state.
— Just-in-time, context specific visual and/or auditory guidance may be provided to field operators through mobile digital devices, for example to help them avoid hazardous zones (e.g. due to fire, flooding or radiation, see Sections 3.12 and 6.8) and perform the right sequences of actions at the right place and on the right equipment, in adequate coordination with other actions performed elsewhere in the installation.

Guidance can be particularly useful when domain experts (e.g. computer security experts) are off-service and immediate action has to be taken by night shift personnel. It may be provided directly by automated tools embedding adequate expertise, but it may also be provided by human experts on-call. Tools then could be used to facilitate the two-way communication between the experts on call (who need information to fully understand the situation) and the local operators (who need unambiguous directives).

5.3.1. Benefits to plant performance

Such guidance can not only help reduce human error, but it can also help optimize the time and/or the amount of material necessary and enhance workers' safety. It may also be an element to consider when organizing the operations team.

5.3.2. Examples

The following are some illustrative examples of decision making aids for operation:

(a) Visual presentation of linings and tagging: A lining error in the preparation of an intervention has led to the introduction of inadequate material in the reactor of a nuclear power plant, which had to be shut down for several months. To prevent such costly errors from ever occurring again, interventions are now planned and prepared for with the aid of tools showing automatically and visually on CAD diagrams the effects of linings and tagging.

(b) Faster-than-real-time simulation: Faster-than-real-time simulation may sometimes be used to verify that a given course of action will indeed achieve what is expected and will not have undesirable side effects.

(c) Use of 3-D models to prepare for the replacement of large components: When a large component, such as a steam generator or a turbine, needs to be installed or replaced, precise preparation is often necessary to determine a sequence of movements ensuring that the component can be put into place or removed without damage. To facilitate such preparation, 3-D models of the component and of the facility have sometimes been used. These models are either CAD models (as designed) or from 3-D scanning (as built).

5.3.3. Maturity

Even though there are already many cases of guidance in the determination of appropriate courses of action, this kind of approach is still largely untapped, considering its huge potential for improvement of plant performance.

5.4. ROBOTS AND DRONES

In some nuclear power plants, the use of autonomous or remotely operated robots and/or drones for inspection, maintenance and repair works has become an integral part of plant operation. Due to their high flexibility, some robots can be used for recurrent but also for non-recurrent tasks. As the precision of positioning, trajectories and actions keeps improving, the scope of use for robots, drones and other remotely operated tools in nuclear power plants continues to expand.

5.4.1. Benefits to plant performance

Robots, drones and other remotely operated tools:

— Can operate in areas unsafe for human workers (e.g. due to radiation or chemical pollution).
— Can carry out interventions in spaces too narrow for, or otherwise inaccessible to, human workers.
— Can perform recurrent tasks repeatedly without fatigue and with high reliability, high precision and high efficiency. That may result in significant cost and time savings, in particular in the case of tasks on the critical path of an outage.

5.4.2. Examples

The following are some illustrative examples of different kinds of robotic and remotely operated tools used for non-destructive examinations in some operating nuclear power plants:

(a) Underwater inspection robots: The underwater, remotely operated inspection robot shown in Fig. 8 has been designed for the non-destructive examination of reactor areas with very high radiation levels. Thanks to its flexibility, ease of use and high manoeuvrability, inspection times have been reduced significantly without compromising accuracy or reliability. Due to its radiation tolerant and resistant components, it has been highly reliable for more than 15 years. With easily attached add-ons (e.g. grippers or suction devices), it can also be used for other purposes, such as searching for and retrieving foreign objects or monitoring the handling of underwater equipment.

(b) Robotic systems for non-destructive reactor pressure vessel examinations: The need for high precision robotic solutions is especially evident for in-service non-destructive examination of the

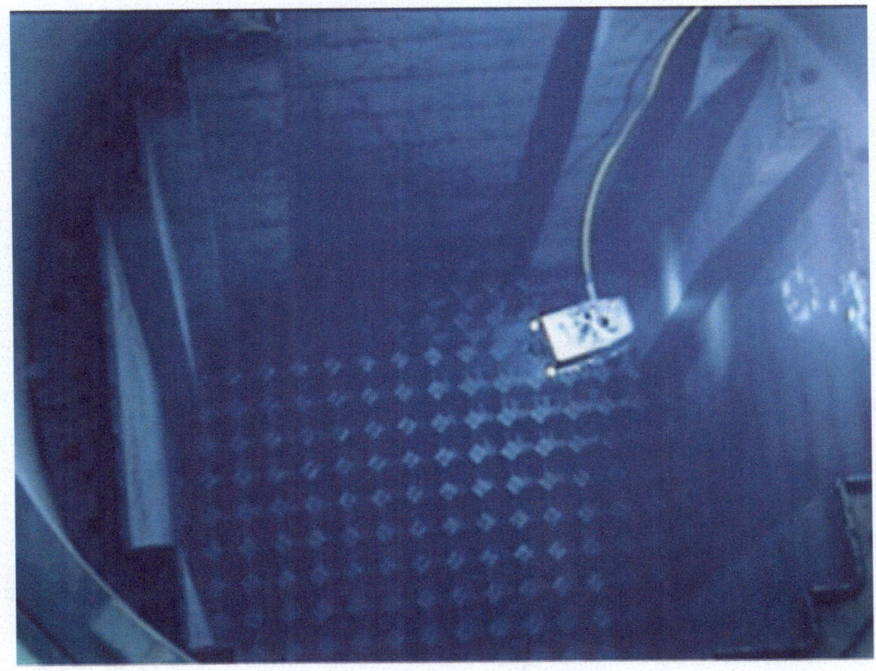

FIG. 8. Underwater robot during a plant outage (courtesy of Framatome, Germany).

reactor pressure vessel and its essential components, such as the primary nozzles and their weld seams to the primary coolant line. With high definition images, eddy current testing and ultrasound testing, the robot can proceed to detailed inspections, looking for any internal cracks or defects. High operating speed may sometimes contribute to shortened outage times.

(c)   Robots for high precision operations in cramped spaces: Robots may also be used to carry out operations in areas that are difficult to access, such as the space between the enclosures of reactor buildings or any other cramped area where precision operations need to be carried out. For example, a remotely operated robot has been used to repair containment penetration welds inside the piping of a secondary steam circuit.

(d)   Drones: Drones (also known as unmanned aircraft systems) may be used for the inspection of elevated and hard to reach structures, avoiding the high cost and long delays of scaffolding. References [74, 75] provide extensive information on the use of drones for nuclear power plants.

5.4.3.   Maturity

Robots and drones are nowadays routinely used in many operating nuclear power plants, and new applications abound.

5.5.   FAULT TOLERANT CONTROL

Fault tolerant control (FTC) is a recently developed approach that works separately from and ahead of safety systems to prevent simple, local faults from developing into failures detrimental to plant availability, performance and safety. Unlike traditional control, it blends several disciplines (including on-line monitoring and diagnostics, automatic condition assessment and automatic determination of remedial actions) and operates as a supervision system that identifies faults' type, location and severity and then adjusts control parameters (fault accommodation) and structure (control reconfiguration/allocation)

TABLE 2. FTC TECHNIQUES

Fault type	FTC technique
Sensor faults	• Physical redundancy • On-line monitoring • Sensor recalibration • Sensor data reconciliation (see Section 3.6)
Controller faults	• FTC itself
Actuator faults	• Adaptive control
Plant faults	• Control reconfiguration or allocation • Supervision or decision making

to maintain nominal plant performance. In some cases, it may have to adjust plant performance objectives (supervision or decision making).

FTC is implemented with ad hoc techniques that can be classified as shown in Table 2.

5.5.1. Benefits to plant performance

FTC aims, in the presence of faults, at maintaining plant availability and performance as best as safety allows and preventing these faults from causing damage and safety hazards. It possibly reduces the use of physical redundancy and complex fault monitoring systems. It increases plant operation reliability while reducing operator workload.

5.5.2. Examples

The following are some illustrative examples of FTC systems implemented in nuclear research reactors:

(a) On-line fault tolerant controller for MITR-II: An experimental digital FTC system has been implemented in a 5 MW nuclear research reactor, MITR-II [76]. It incorporates on-line detection and reconfiguration of faulty equipment, sensor calibration and information display within a structure of alternatively configured regulators for plant control. The system was found to be tolerant of one or two failures in power, flow and temperature sensors. It is capable of calibrating the power sensors on-line and maintaining the reactor power at the desired level during both steady state and rapid transient conditions.

(b) Reconfigurable hybrid system for a feedwater system of the Experimental Breeder Reactor II (EBR-II): A supervisory control system with a reconfigurable hybrid architecture that selects a suitable set of active control and supervisory algorithm based on the process status has been experimentally implemented on the feedwater system at the EBR-II of Argonne National Laboratory. Performance of the hybrid system to accommodate component failures was evaluated through in-plant experiments involving controller gain verification and tuning of the alternative pressure based condensate controller and reduction in the steam flow to the heater [77].

5.5.3. Maturity

Some FTC techniques are mature in the nuclear industry, for example process data reconciliation (see Section 3.6) and improved diagnostics and prognostics (see Section 4). Mode switching control with limitation functions is already in use in some operating nuclear power plants. On-line sensor recalibration is also used in some plants, although not under the name of FTC. Reconfigurable or reconstructible control has been studied in academia but not applied in the nuclear industry. Control allocation is not completely mature yet in the nuclear industry.

5.6. REMOTE SUPPORT

Some nuclear power plant fleet operators have set up a remote support centre gathering their best experts to provide support to local teams in case of accident and emergency situations. However, as is increasingly done for non-nuclear power plants, such centres could also provide expertise for normal operation with, for example, support for:

— Planning, management and decision making for day-to-day operation and outages.
— Plant, plant systems and equipment condition monitoring (see Section 4.1.1).
— Incident management, including management of computer security events.
— Management of a unified, fleetwide information and data repository.
— Collection and archival of plants' operational data and experience and big data analyses.
— Management of technical documentation.
— Predictive analytics for:
  • Assessment of equipment condition and failure probability;
  • Diagnostics and prognostics;
  • Condition based maintenance.
— Control of work performance.
— Management of workers' safety.
— VR or AR.
— Control of autonomous equipment in high hazard zones.

With high levels of data communication between such remote support centres and plants, computer security needs, of course, to be ensured.

5.6.1. Benefits to plant performance

With this approach, it is possible to mutualize the cost of expertise across a fleet and improve operational efficiency.

5.6.2. Examples

The following are some illustrative examples of remote support solutions for plant operation and maintenance:

(a) NuScale: Because a NuScale SMR plant can be built in remote locations, it may have limited on-site resources and tough conditions. Advanced monitoring and diagnostic capabilities can be employed in a remote support centre if associated computer security concerns are addressed. The NuScale system design can also provide a central repository to acquire, archive and process all modules' operating and maintenance data to enable remote condition monitoring and predicative maintenance solutions.
(b) Rosenergoatom: Based on a number of pilot nuclear power plants (at Novovoronezh, Kursk and Smolensk), an operation support information centre was created. Its unified information platform collects operational measurement data and data on various objects, including digital models, to

support operation, maintenance, engineering, resource management and documentation. It will be integrated with other Rosenergoatom information systems.

(c)   Remote outage control centre: During an outage, a reactor internal element was found to be displaced. After intensive communication between the local team and experts at the remote outage control centre, a repair plan based on a bespoke tool and a bespoke repair process was developed. It was first tested on the mock-ups of the remote centre and then implemented by the local team entirely by itself. Planning, testing and actual repair did not affect outage duration. Power production resumed as initially planned.

(d)   Remote non-destructive testing: Traditional non-destructive testing activities require the on-site presence of cohorts, operators and accredited evaluators. Operators were confronted with travel restrictions but could perform non-destructive testing remotely through preconfigured, robust virtual private network (VPN) connections. These connections were deployed within two days, a few weeks before outage, and lasted throughout the plant outage. The original outage schedule was maintained. While VPNs are not a novelty, the urgency made decision making possible, providing the industry with references that will allow for more flexibility in unprecedented situations during future outages.

5.6.3. Maturity

A number of remote support centres for accident and emergency situations are already in service in nuclear power plant fleets, but although remote support centres for day-to-day operation are common in non-nuclear fleets, they are still rare in nuclear power plant fleets. However, many SMR programmes feature such centres.

5.7. COMPUTER SECURITY MONITORING AND INCIDENT MANAGEMENT SYSTEMS

Digital technologies are pervasive in nuclear power plants and are used in diverse applications, including I&C systems, physical protection systems, smart equipment and devices used to collect information on plant performance or for radiation protection, and computer based aids that support operations and maintenance tasks. These technologies are susceptible to cyber-attack, and computer security is consequently an important aspect that needs to be addressed throughout the lifetime of a nuclear facility. Specific computer security guidance for nuclear facilities can be found in Refs [12–14], among others.

It is possible to assess computer security events by auditing digital systems configurations to identify unauthorized changes or by monitoring their network traffic to identify unexpected traffic or patterns known to be associated with cyber-attacks. However, it is not practical for humans to do this when a facility may have thousands of computers and a high volume of network traffic. Given the complexity of monitoring and analysing security related data in modern computer based systems, applications and tools can assist in the management and analysis of security related data.

Security information and event management (SIEM) and security orchestration, automation and response (SOAR) are examples of tools that support the detection and the diagnosis of events that could be due to malicious cyber-attacks and then provide guidance in the defence against such attacks. These tools have been applied to monitor a variety of data sources to detect possible computer security events. They collect data from various sources and of various types, such as observation data on digital systems, equipment, devices and communication links; data from physical security management systems (e.g. access control to computer security zones, audio and/or video streams from plant monitoring devices); data from work management systems (e.g. when interventions are scheduled in computer security zones and by whom); and data from computer security monitoring systems of other plants. They have the capabilities to normalize, aggregate and detect anomalies and notify parties when suspicious behaviour is detected.

SIEM and SOAR are mainly used for information technology systems and are less common in operational technology systems. The main function of SIEM is to provide an alert of a possible computer security event. SIEM typically has the following features:

— Collection of data from various data sources, such as system security event logs.
— Normalization, correlation and aggregation of these data.
— Application of rules to detect suspicious behaviours. Examples of simple rules could be if a user login failure occurs more than a certain number of times within a specified period, or if an antivirus program detects a virus.
— Alerts that can be received by security staff.

SIEM is very widely used in the field of information technology to detect cybersecurity events, but there are some challenges. For instance, labour is needed to tune the rules to reduce the false positive rate, and there is a need for skilled staff to perform analysis activities and act. This is labour intensive and can lead to staff fatigue. SOAR — which generally refers to technologies that enable handling security processes including threat and vulnerability management, incident response and security operations automation — addresses these shortfalls; SOAR typically:

— Performs all functions of SIEM (SIEM solutions are part of the SOAR platform).
— Includes or may include automated functionalities, such as:
  • Triage.
  • Data collation based on playbooks or user request.
  • Data aggregation and normalization.
  • Use of human inputs to respond to threats or automatic response to threats.
  • Threat identification.
  • It may use 'playbooks' to act.
  • Case management capabilities.
  • Integration with external security tools.
  • AI and machine learning.

5.7.1. Benefits to plant performance

SIEM and SOAR are potentially useful to support computer security at nuclear facilities. Care has to be taken when considering them for I&C systems performing functions important to nuclear safety, since they could potentially impact safety function.

Events that could be due to malicious attacks against computer security can potentially occur at any time. Computer security requires specialized technical knowledge and experience and, in nuclear power plants, it also requires operational technology expertise. Maintaining a continuous presence on a nuclear power plant site of highly skilled and often highly paid computer security experts may be costly. However, failure to detect and correctly analyse such events could result in successful attacks with possibly significant consequences, including loss of performance.

By aggregating and assessing information from various data sources, efficient SIEM and SOAR tools may help in deciding the best defensive or mitigation action, guiding the staff trained in responding to the incident or delivering an automated response. It may also provide off-site experts (e.g. at a remote support centre, see Section 5.6) with information allowing the plant personnel to carry out precise diagnostics and decide on the next action.

5.7.2. Examples

Examples of open source integrated computer monitoring systems are AlienVault OSSIM [78] and Security Onion [79]. In addition, research is currently being undertaken, such as the Canada-US Blended

Cyber-Physical Exercise conducted by Canadian Nuclear Laboratories and Sandia National Laboratories, on a method for assessing computer security incident response actions, which includes evaluation of the contribution that integrated security management systems can make during an incident response.

5.8. ON-LINE OPERATOR DOCUMENTATION

In addition to paper based operation manuals, many nuclear power plants have electronic operation manuals, with information specifically relevant to the current plant state or to the current activity made directly available to plant operators, right from their computerized workstations, in the form of on-line documentation or even step-by-step guidance.

5.8.1. Benefits to plant performance

Compared to conventional, paper based operation manuals, electronic operation manuals offer a wide variety of benefits and advantages for plant performance, in particular a significant reduction in human error.

Another advance in the reduction of human error is the introduction of modular, decision tree organized operating procedures for improved oversight and understanding of complex parallel operating tasks, which provide:

— Faster retrieval of desired information, in particular when electronic operation manual access is integrated into operation workstations;
— Improved overall consistency, with automated verification of features, such as alarm set points and component designations, via a central engineering database;
— Guaranteed consistency between documentation on-site and documentation at remote support centres;
— Facilitated documentation management across a fleet;
— Simple and consistent generation of paper based documents.

5.8.2. Example

One illustrative example of on-line operator documentation for nuclear power plants is electronic operation manuals for EPR reactors. Extensive testing on full scale plant simulators has led to further evolution of the electronic operation manual concept in EPR plants, in particular regarding HFE, human–machine interfaces and plant specific adjustments (see Refs [72, 80]).

5.8.3. Maturity

Electronic operation manuals have been in use in nuclear power plants for many decades. However, although they offer many benefits, they require rigour and supporting tools.

5.9. WORKFLOW OPTIMIZATION

Process optimization aims at improving the overall efficiency of the numerous technical processes required for the operation of a nuclear power plant, within the limits set by safety, security, regulations and overall plant constraints. It may follow steps, such as:

— Careful selection of the key performance indicators to be taken into consideration;

— Identification of processes that could be or need to be improved, according to the key performance indicators;
— Critical analysis and reconsideration of each identified process's purpose within the overall workflow and the role, added value and effectiveness of its constituent activities;
— Definition of new or optimized, more effective processes, taking into account possible needs for training and/or technical infrastructures (e.g. documentation, tools or databases);
— Implementation of new processes;
— Monitoring of the new processes, evaluation against the key performance indicators and possibly introduction of improvements.

Various process optimization methods have been developed, such as:

— The PDSA (plan, do, study and act)/STAR (stop, think, act and review) method;
— The process mining method;
— The DMAIC (define, measure, analyse, improve and control) method;
— The Kaizen method.

5.9.1. Benefits

Well thought out process optimization can provide benefits at many different levels, including economic benefits, safety improvements, human resource management improvements and environmental benefits.

5.9.2. Examples

For examples of how digital tools can support the effectiveness of engineering processes in nuclear power plants, see Annex VI. See also Ref. [81] for how work management can increase worker productivity and reduce operating costs.

5.9.3. Maturity

Workflow optimization has been successfully implemented in some nuclear power plants, as illustrated in Annex VI. However, in many nuclear power plants, there is still considerable scope for further progress.

5.10. SUPPORT FOR PLANT LONG TERM OPERATION

Long term operation is a key point of the profitability of the plant, since poor long term operation may lead to shortened life or to colossal refurbishment costs. Successful long term operation depends on the ability to preserve both skills and equipment all along a plant's life cycle.

5.10.1. Benefits

There are many ways in which advanced digital technologies and digital I&C can help improve the longevity of plants:

— They can improve knowledge of the environment and constraints really experienced by equipment, for example, with temperature, humidity, vibration and other operating conditions sensors (such sensors exist for electronic boards, for instance). This improved knowledge can be used to detect early signs of degradation, to improve maintenance programmes and reduce the stress caused by interventions and to better estimate the RUL of equipment. Such monitoring can also be extended to passive and/or non-replaceable components. With innovative sensing methods, low power technologies and wireless data communication, sensors monitoring key ageing parameters can now operate many years on batteries and can be placed in locations that were not anticipated by the initial plant designers.

— Advanced digital controls can be used to reduce equipment constraints. This is often seen with smart actuators, where embedded controllers incorporate expertise accumulated over decades of experience across multiple industries. Similarly, I&C limitation functions can prevent some of the reactor trips and thus avoid stressful reactor conditions that shorten a plant's lifetime.
— Knowledge management is a key point of long term operation (see Section 3.13).
— Advanced technologies can play a role in limiting human error (see Section 6.7), the most severe consequences of which could affect a plant's lifetime.

5.10.2. Maturity

The principle of environmental sensors allowing continuous monitoring is well known but not sufficiently used in plants. The other points mentioned above are, for the most part, still under development.

6. SUPPORTING TECHNOLOGIES

Beyond the approaches outlined in the previous sections, a variety of supporting technologies and related approaches can provide advanced functional and operational capabilities potentially beneficial to plant performance, as discussed in this section. Digital I&C technologies such as smart devices, wireless data communication and multimodal sensors can improve data acquisition, enhance system monitoring and enable more efficient control, and adherence to industrial standards and products for I&C helps ensure cost effectiveness, reliability and interoperability. Simplification of I&C systems and reduction of the number of I&C platforms further contribute to streamlined architectures, while advanced human–system interfaces reduce the potential for human error and improve operators' effectiveness. Additional technologies such as mobile digital assistance, virtual and augmented reality, and 3-D printing offer new capabilities for maintenance, training and construction. Integration of construction data and common information models supports consistency across the plant life cycle, while big data analytics and artificial intelligence provide powerful tools for further optimizing performance and decision making.

6.1. DIGITAL INSTRUMENTATION AND CONTROL TECHNOLOGIES — SMART DEVICES

Smart devices are pieces of equipment that contain and are controlled by software or programmed logic, so that they can provide services and offer features, such as:

— Advanced functional and operational capabilities;
— Easy adaptability with predefined, configurable parameters;
— Self-monitoring and embedded diagnostics and prognostics;
— Multiplexed data communication;
— Interfaces compliant with well recognized industrial standards;
— Certified compliance with safety, computer security and/or interoperability standards;
— Vast amounts of operational experience, generally in large part from non-nuclear industries.

In practice, for certain types of devices, it is increasingly difficult and sometimes outright impossible to find non-smart products. However, the use of smart devices in applications important to safety raises a certain number of issues (see Ref. [82] for safety aspects and criteria associated with the safe use of industrial commercial smart devices in systems important to safety).

6.1.1. Benefits to plant performance

Though there are concerns, such as computer security, electromagnetic compatibility, obsolescence and, sometimes, safety and common cause failure (in the case of use in applications important to safety, see Refs [82, 83]), smart devices generally offer many characteristics potentially beneficial to plant performance:

— Advanced functional and operational capabilities may benefit plant processes' efficiency and therefore power production.
— Easy adaptability enables the use of predeveloped, off-the-shelf devices, which are generally less costly to purchase and operate than bespoke devices.
— Self-monitoring and embedded diagnostics and prognostics improve reliability and facilitate maintenance optimization.
— Multiplexed data communication can significantly reduce cabling costs (installation, verification and replacement).
— Certified interface compliance with well recognized industrial standards facilitates interoperability between smart devices and I&C systems from different suppliers and replacement in case of obsolescence.
— Certified compliance with safety and/or computer security standards can facilitate licensing.
— Vast amounts of operational experience generally allow a gradual removal of design errors and can provide reliability figures for probabilistic studies.

6.1.2. Examples

The following are some illustrative examples of smart I&C devices:

(a) Valve positioners: Several vendors propose smart valve positioners with self-monitoring capabilities such as:
    (i) Periodic slight movements that do not affect the function of the valve but that help verify that the valve is not stuck;
    (ii) Measurement of the force that needs to be applied to move the valve: Deviations from nominal values could be harbingers of upcoming seize ups.
(b) Smart sensors: Measurement accuracy is often highly dependent on ambient and process conditions (e.g. temperature). Smart sensors improve accuracy by taking account of these conditions to perform necessary corrections in situ and in real time. With reduced process measurement uncertainty, the plant can operate close to its safety limits and hence produce more power.
(c) Field buses and remote input/output (I/O): The newly built Hualong One nuclear power plant uses fieldbuses and remote I/O for its non-safety digital control system to improve reliability and reduce costs.

6.1.3. Maturity

Smart devices have been in extensive use in non-nuclear industries for many decades. Some of these industries, such as the process, petrochemical or manufacturing industries, need to address safety and computer security requirements comparable to those that the nuclear industry requires.

Smart devices are also in extensive use in the nuclear industry but often not at their full potential (e.g. they are often installed with point-to-point connections, and their self-monitoring, self-diagnostics and self-prognostics capabilities are sometimes discarded).

6.2. DIGITAL INSTRUMENTATION AND CONTROL TECHNOLOGIES — WIRELESS DATA COMMUNICATION

Data communication plays a vital role in safe, secure, reliable and efficient operation. In nuclear power plants, it is mainly based on hardwired point-to-point or multiplexed cables. However, cabling is notoriously expensive, particularly after the initial build, when it becomes necessary to add or improve permanent instrumentation, replace existing cables or put in place temporary instrumentation to support specific activities. Although wired communication will remain the norm for some applications (in particular applications important to safety), nuclear power plants are also exploring wireless communication (see Ref. [54]). Reference [84] presents a three tier strategy supporting wireless transmission of in-core measurements to the control room or a secure cloud platform for control, analytics and decision making purposes.

6.2.1. Benefits to plant performance

When applicable, wireless communication presents several benefits, including:

— Significant cost savings, as initial cabling and cabling replacement are both expensive and time consuming [85];
— Opportunities for plant performance improvements that would not be cost effective or would be too complex to implement with wired communication [86];
— Improved coordination of workers during operation and outages, with benefits in terms of efficiency, security and worker safety, including reduced exposure to radiation;
— Access to data that were previously unavailable [86].

6.2.2. Examples

The following are some illustrative examples of applications of wireless data communication:

(a) Instrumentation:
    (i) Wireless vibration sensors' installation on a circulating water system to achieve on-line measurement that was previously collected at periodic intervals manually [84];
    (ii) On-line valve verification using wireless communication to achieve efficiency and intelligent plant configuration, which was initially performed manually [87];
    (iii) A wireless pressure sensor for high temperature applications in the nuclear environment [88].
(b) Mobile digital assistance: See Section 6.8.
(c) Voice communication: The Comanche Peak Nuclear Power Plant installed a wireless network throughout the entire plant for their voice communication purposes. Also, the Diablo Canyon Power Plant uses wireless dosimetry and a wireless paging system in the plant [89].
(d) Condition based monitoring: Wireless sensors and data transmission equipment has been installed to implement condition based maintenance without installing costly cable intensive sensors [90].

6.2.3. Maturity

Wireless communication is still in the early stages of implementation in nuclear power plants, due to the following challenges:

— Concerns related to computer security;
— Concerns related to electromagnetic interference and radiofrequency interference;
— The need to develop multiband heterogeneous architecture to support a wide range of frequency ranges (see Ref. [91]).

6.3. DIGITAL INSTRUMENTATION AND CONTROL TECHNOLOGIES — MULTIMODAL SENSORS

In the context of this publication, a multimodal sensor is a single sensor capable of simultaneous measurements of more than one quantity.

6.3.1. Benefits to plant performance

Multimodal sensors are expected to be especially important for condition based maintenance in nuclear power plants by enabling greater sensor deployment density and measurements of key quantities at necessary points within a nuclear power plant. It is expected that such sensors will also eventually be deployed as the technology matures, given the advantages of monitoring multiple quantities using a limited set of sensing wells/penetrations.

6.3.2. Examples

The following are some illustrative examples of multimodal sensors used to support condition based maintenance:

(a) A piezoresistive sensor, as described in Ref. [92], can simultaneously measure normal and shear force along with temperature.
(b) A biaxial vibration sensor, as described in Ref. [84], can simultaneously measure vibration and temperature.

6.3.3. Maturity

There are some challenges with the application of multimodal sensors, many of which arise from the cross-sensitivity of each measurement to others, as a single sensing element is often used. Among these challenges are:

— The need to quantify the uncertainty and cross-sensitivity in each measurement;
— The lack of qualification methods and associated codes and standards for multimodal sensors;
— The reliability of multimodal sensors and the impact of failure in any individual (or more than one) measurement mode;
— Instrumentation design and integration with multimodal sensors.

6.4. DIGITAL INSTRUMENTATION AND CONTROL TECHNOLOGIES — INDUSTRIAL STANDARDS AND PRODUCTS

The nuclear industry represents a limited I&C market, and nuclear specific I&C products tend to be expensive and benefit from only limited experience in operation.

6.4.1. Benefits to plant performance

The use of appropriate industrial standards and products from the general industry for I&C functions of low or no safety importance brings many benefits, in particular:

— Cost effectiveness, since such products benefit from competition in a much wider market than the sole nuclear industry. Also, as they are more likely to be based on up-to-date, high performance digital technologies, less hardware is needed, with many beneficial effects: lower power consumption and lower heat dissipation mean lower demands on support systems (power supplies and HVAC), lower demands on civil engineering and lower demands in terms of installation on-site and maintenance.
— Safety does not concern only the nuclear industry; it also concerns many other industrial sectors, such as the petrochemical and manufacturing industries, and it is at the heart of standards like IEC 61508 [93] and IEC 61511 [94]. Even though the rigour of their safety requirements might not be sufficient for I&C functions and systems of high importance to nuclear power plant safety, these standards often offer a good basis for I&C functions and systems of low or no safety importance. In addition, many industrial products come with a certification of compliance.
— Computer security, like safety, does not concern only the nuclear industry; it concerns all industrial sectors, and it is at the heart of industrial standards like the OPC Foundation's Unified Architecture standard (OPC UA) [95].
— Interoperability between I&C systems from different vendors, with standards like OPC UA. This is often a significant asset when I&C systems embedded in plant components (e.g. a turbine or a water chilling system) need to be interfaced with the plant I&C systems.
— Portability of applications, with standards like IEC 61131-3 [96] and guidelines from the PLCopen organization [97] to facilitate the replacement of I&C systems that have become obsolete.

6.4.2. Examples

Figure 9 shows a chemical plant where each instrument and control system supports the OPC UA protocol. I&C devices and systems, audio and video systems, and management and business systems form a unified integrated whole where each device and system can provide real time data, historical data, event records and other information to other systems and where big data analytics (see Section 6.13) can be fruitfully applied.

6.4.3. Maturity

I&C products from the general industry have been used in nuclear power plants for many years, but OPC UA is a relatively recent standard. Also, portability of I&C applications with PLCopen is not a widespread practice.

6.5. DIGITAL INSTRUMENTATION AND CONTROL TECHNOLOGIES — SIMPLIFICATION OF SYSTEMS

Nowadays, new or modernized safety I&C systems are generally digital. That brings a number of important benefits to plant performance, in particular with the ability to reduce measurement uncertainties

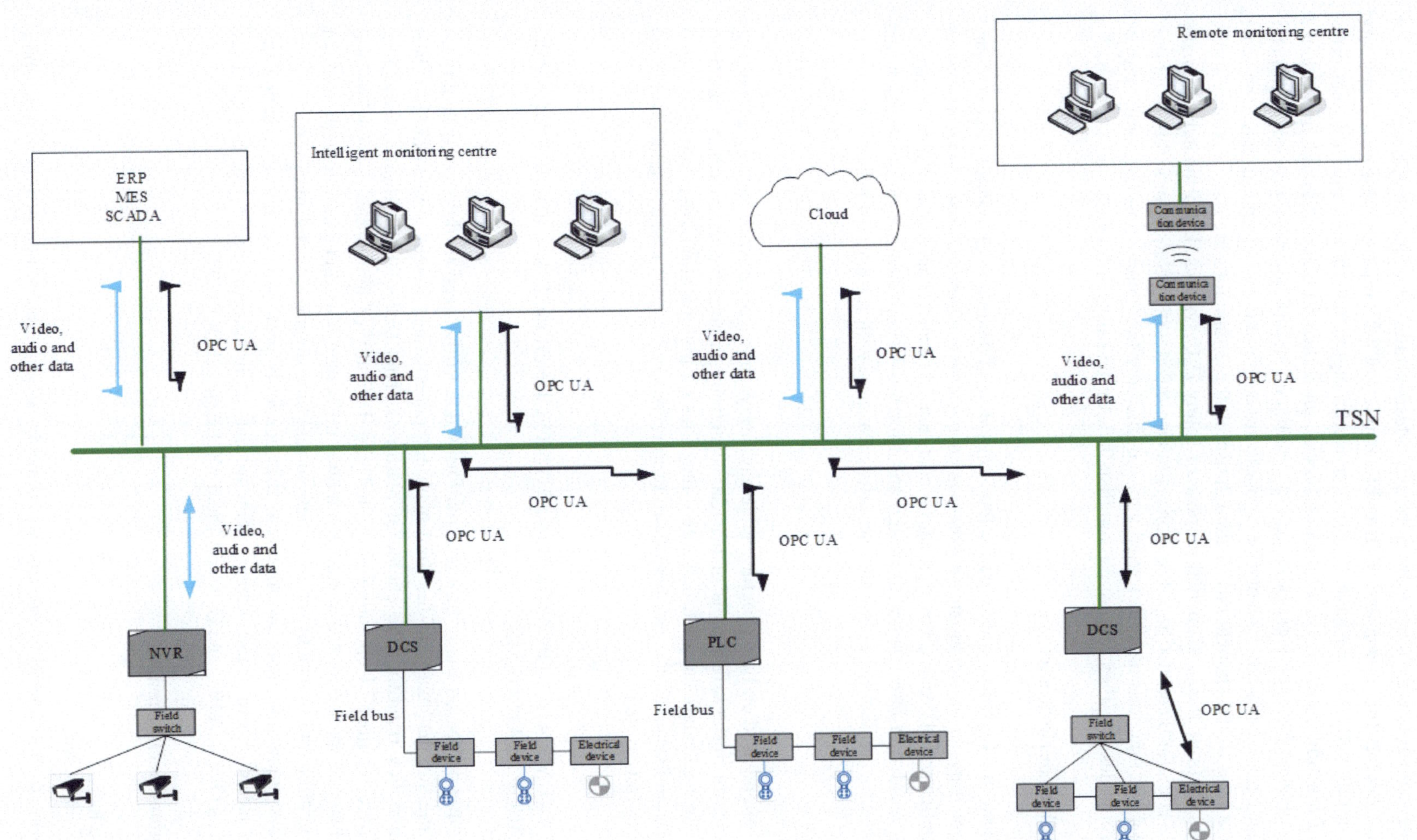

FIG. 9. *Example of a unified integrated network architecture. DCS — distributed control system; ERP — enterprise resource planning; MES — manufacturing execution system; NVR — network video recorder; PLC — programmable logic controller; SCADA — supervisory control and data acquisition; TNS — terminal network server. (Courtesy of Y. Li, China Nuclear Power Engineering Corporation, China.)*

and to better exploit reactors within their safety margins. The first generation of digital safety I&C systems (in the 1980s) were relatively simple due to bespoke design (tailored to the specific needs of each reactor) and the limited computing power and addressable memory of 8-bit microprocessors. Complexity has since increased significantly with the advent of more powerful digital technologies and I&C platforms offering a number of functions, such as basic function blocks, input data validation through coincidence checking, self-surveillance, facilitated periodic testing, facilitated modification of process parameter values, aids to debugging and automatic adaptation to hardware configuration (in order to have the same system software regardless of the application). Although useful, these functions are only auxiliary to the safety functions that constitute the fundamental purpose of safety I&C systems. Unfortunately, in many designs, the software code for these auxiliary functions is intermingled with the code for safety functions, to the point that auxiliary functions have increased very significantly the volume and the complexity of safety software (in some designs, they represent more than 95% of the safety software).

6.5.1. Benefits to plant performance

The larger volume and higher complexity of safety software and the increased quantity of necessary hardware have severe impacts on development costs, licensing costs, hardware costs, support system costs (i.e. power supplies and HVAC), civil engineering costs and, ultimately, operation and maintenance costs, which can be avoided in large part with design and functional simplification. Simplification and returning to basics tend to also facilitate computer security measures, with further cost reduction.

6.5.2. Examples

Some designs, often based on field programmable gate arrays, have aimed with success at avoiding that additional complexity in the safety parts of the I&C system by segregating the auxiliary functions into parts that, in hardware design, cannot affect safety functions during operation. Also, with the use of passive safety features and systems, the number of safety functions can be decreased, leading to further simplification of safety I&C systems.

6.6. DIGITAL INSTRUMENTATION AND CONTROL TECHNOLOGIES — REDUCTION OF THE NUMBER OF PLATFORMS

I&C is often implemented on many different platforms due to the need of defence in depth and diversity but also to the fact that many off-the-shelf plant components and systems embed I&C of their own with no general policy regarding the choice of platforms. This lack of unification significantly increases the knowledge required of operating personnel, the difficulty of integration and testing, and the quantity and diversity of spare parts. As it can also reduce interoperability and data integration, it can limit the level of automation and reduce the effectiveness of real time diagnostics and analyses.

For digital I&C systems not important to safety, it is generally preferable to aim at reducing the number of platforms used and also to use industry standard data communication protocols designed for interoperability, such as OPC UA [95] and OPC UA Field eXchange (UAFX) [98].

6.6.1. Benefits to plant performance

The approach reduces:

— The need for protocol converters that are costly to develop, acquire, install and maintain in the long term;
— The number of I&C platforms to be qualified and that will need to be replaced in the long term;

— Training costs, with fewer I&C platforms on which operation and maintenance personnel need to be trained;
— Maintenance costs, with fewer long term maintenance contracts with suppliers;
— Spare parts management costs.

6.6.2. Maturity

The industrial standards cited are relatively recent, but major industrial organizations have placed them at the centre of their digital transformation strategy (see Ref. [99]).

6.7. DIGITAL INSTRUMENTATION AND CONTROL TECHNOLOGIES — ADVANCED HUMAN–SYSTEM INTERFACES

Digital HSIs are nowadays ubiquitous in the industry and in everyday life. With appropriate design and HFE (see Refs [16, 100, 101]), they can provide operators (in the control room and in the field) with a clear view of the current situation and all necessary information, guidance and support for performing what is required of them at any given instant.

6.7.1. Benefits to plant performance

Advanced digital HSIs can contribute to plant performance by:

— Reducing the potential for human error that could affect safety, security, plant performance and workers' safety;
— Enhancing operators' effectiveness, with reduced cognitive effort and workload.

6.7.2. Examples

The following are some illustrative examples of advanced HSI technologies for plant operation and monitoring.

6.7.2.1. Visual presentation of linings and tagging

See Section 5.3.2(a).

6.7.2.2. Computer based procedures

Computer based procedures guide operators in the application of procedures, taking account of the state of the process and of plant equipment (see Refs [71–73]).

6.7.2.3. On-line operator documentation

See Section 5.8.

6.7.2.4. Augmented reality

See Section 6.9.

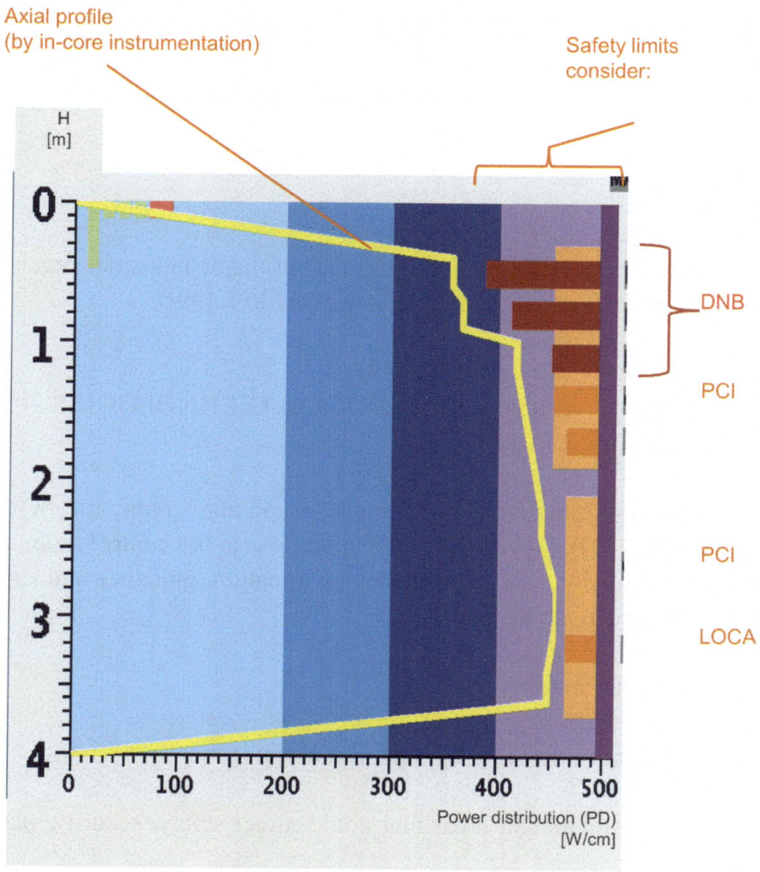

FIG. 10. Visualization of the axial power distribution and safety limits (adapted from Ref. [24]).

6.7.2.5. Visualization to support plant operation

The visualization of the reactor state from the safety I&C plays an important role for operators. Figure 10 [24] presents a typical core visualization on German PWRs. It shows the shape of the axial power distribution (yellow curve) based on the results of in-core measurements, together with the safety limits (orange and red bars), such as departure from nucleate boiling (DNB), PCI and LOCA.

The position of the control rods is also seen in the upper left corner (green and orange bars). This coherent representation of rod positions and the power distribution margins allows operators to track the actions of the automatic controller and assess whether power can be further increased.

Introducing digital I&C for reactor control, the visualization was further developed in cooperation with utilities to include prediction technology (see Ref. [24]) and thus fulfil the World Association of Nuclear Operators requirement for plausible reactivity management during flexible operation in a suitable and automated way.

6.7.2.6. Graphical core monitoring interface

A CMS provides operators with detailed power 3-D density and burnup distributions in the core and other information about the status of the core (e.g. measured and calculated power margins, see Ref. [39]). In state of the art CMSs, this information is presented through a configurable graphical user interface. Displays can be stored and adjusted according to operational needs or individual preferences. Along with a high degree of automation (recurring analyses and reports), operators consequently benefit from time savings and error reduction.

6.7.3. Maturity

Digital HSIs have been in use in many nuclear power plants for non-safety and even safety related applications for decades now, and they feature in most, if not all, new builds. Some nuclear power plants also have them for safety applications in a more limited manner due to licensing constraints and to defence in depth and diversity requirements.

6.8. MOBILE DIGITAL ASSISTANCE

A mobile digital assistant is a handheld or wearable device (e.g. a smartphone, an electronic tablet or augmented realty goggles; see Section 6.9) that can provide a variety of services to mobile staff (e.g. field operators, construction and installation workers, fire fighters and rescue teams), such as:

— Audio, video and/or data communication with control room operators or other mobile staff;
— Localization and navigation within the plant to find the quickest path to a destination while avoiding hazardous areas;
— Video and audio interfaces for aiding in the performance of specific tasks;
— Plant equipment identification, ensuring that actions are performed on the right pieces of equipment and also providing status information on these pieces of equipment;
— Continuous monitoring of ambient conditions (e.g. radiation levels);
— On the fly generation of intervention reports.

6.8.1. Benefits to plant performance

Mobile digital aids may provide many benefits to plant performance:

— Reduction of human error in the field;
— Improved efficiency in the performance of actions in the field;
— Improved workers' safety.

6.8.2. Examples

The following are some illustrative examples of mobile digital assistance applications:

(a) Visual display of hazardous areas: Mobile digital devices can be equipped with applications informing workers in real time of hazardous areas to be avoided.
(b) Mounting/dismounting instructions: The use of CAD/CAM tools is a widespread practice in the industry, in particular in the design and manufacturing of mechanical pieces of equipment. Many such tools can generate visual and animated instructions for mounting and dismounting the pieces of equipment being designed. Mobile applications can be used to provide these instructions to workers in the field, in a convenient form when and where they are needed.
(c) Process optimization: With mobile support for isolation/tagout, workers do not need to follow complex steps and journey between a main control room and equipment in order to hand over the equipment. It means that time can be saved, with a positive impact on workers' safety and on outage duration.

6.8.3. Maturity

Mobile digital assistants are nowadays ubiquitous in the consumer market and in everyday life. Their use in industry in general and in the nuclear industry in particular is still limited, but rapid growth is expected according to various market studies.

6.9. VIRTUAL REALITY AND AUGMENTED REALITY

In computing, 'virtual' means 'not physically existing as such but made by software to appear to do so'. VR uses computers to create virtual 3-D environments (supposedly representative of some real, physical ones) within which users are immersed and able to interact. As defined in Ref. [100], "virtual reality has a high level of immersiveness, fidelity of information representation and degree of active learner participation compared to other forms of mixed reality."

There are two main VR techniques. Head mounted displays, as illustrated in Fig. 11, immerse their users in a virtual world by displaying stereoscopic images and supporting real time positional and rotational head tracking.

The second technique is the cave automatic virtual environment (CAVE), in which images are projected onto between two and six of the walls of a room sized cube, as shown in Fig. 12. Users can see objects apparently floating in the air and walk around them, getting a proper view of what they would look like in reality.

VR may incorporate a combination of supporting devices and functionalities, such as flysticks, full body tracking, gesture recognition, hand tracking, speech recognition, haptics and omnidirectional treadmills.

As defined by Ref. [102], AR is "an interactive experience of a real world environment where the objects that reside in the real world are enhanced by computer-generated perceptual information" through visual, auditory, haptic, somatosensory and olfactory channels. The difference between AR and VR is that AR alters the user's perception of a real world environment, whereas VR completely replaces the user's real world environment with a simulated one (e.g. Fig. 13).

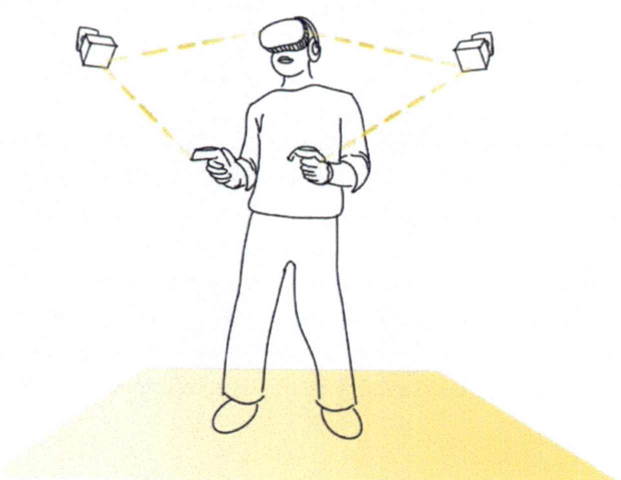

FIG. 11. Virtual reality with head mounted display (courtesy of F. Song, China Nuclear Power Engineering Corporation, China).

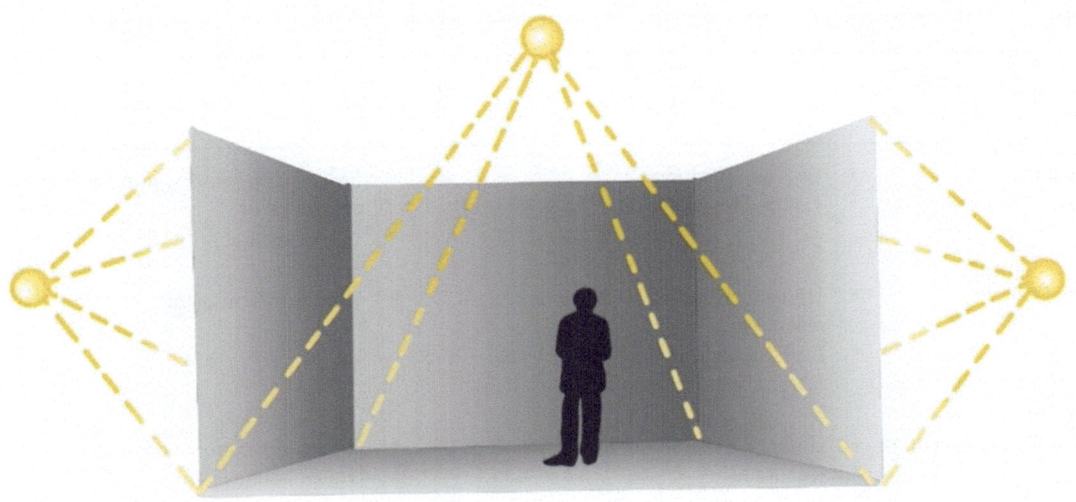

FIG. 12. *CAVE automatic virtual environment (courtesy of F. Song, China Nuclear Power Engineering Corporation, China).*

FIG. 13. *Virtual reality display showing the rotational speed trend of a pump (courtesy of F. Song, China Nuclear Power Engineering Corporation, China).*

Real time object recognition and tracking ensure that real world objects and their associated information are properly aligned with a user's movements and changes of perspective. Table 3 compares sensor based, tag based, physical feature based and mixed recognition and tracking techniques.

AR is mostly useful to mobile workers and is provided through devices, such as:

— Head-mounted devices;
— Handheld devices (e.g. mobile phones, digital notepads, portable computers);
— Holographic projection devices.

TABLE 3. COMPARISON OF AR OBJECT RECOGNITION AND TRACKING TECHNOLOGIES

Object recognition and tracking technology	Strengths	Weaknesses
Sensor based	• Less computation • Lower latency	• Reliance on sensors, influenced by environment • Lower accuracy
Tag based	• Higher accuracy than the sensor based technique • Less computation than the physical feature based technique	• Object drift problem
Physical feature based	• Higher accuracy than the tag based technique • High immersion sensation	• High computational complexity • Higher latency
Mixed	• Suitable computation and accuracy	• Large amount of development • High cost

6.9.1. Benefits to plant performance

Although both require investment and trained personnel, VR and AR can be of great benefit to plant productivity, for the following reasons:

— VR enables activities that, in the real environment:
  • Do not exist yet, in particular in early phases of engineering (see the example in Section 6.9.2.1);
  • Are difficult to replicate in the real world (see the example in Section 6.9.2.2);
  • Are unavailable because equipment is already in use for some other purposes (see the example in Section 6.9.2.3);
  • Are in a too distant location (e.g. for technical experts at a remote support centre, see Section 5.6).
— VR may be used to prepare and verify procedures to be applied in the field during operation, outages or decommissioning, for example to ensure that enough room is available for intervention personnel to reach, examine and use their tools on equipment that they need to operate.
— It may also be used to determine a possible course of action for tasks, such as the moving of large, heavy plant components in restricted volumes.
— It may also be used for training and rehearsal of complex, difficult and/or critical tasks.
— AR may enhance the efficiency of mobile workers, first by helping them locate and identify the pieces of equipment upon which they need to operate.
— It may also provide them with up-to-date information about these pieces of equipment and with real time guidance along the performance of complex, difficult and/or critical tasks, thus reducing the risk of errors that could be very detrimental to plant performance and safety.
— It may improve their safety by providing them with up-to-date information about their physical environment and by helping them avoid contaminated or dangerous areas (e.g. due to radiation, chemical contamination, fire, flooding or earthquakes) and find the best path to where they need to go.

6.9.2. Examples

The following are some illustrative examples of applications of VR and AR in nuclear power plants.

6.9.2.1. *Early human factors verification and validation*

With traditional approaches, human factors verification and validation is based on drawings and data lists, and many defects are identified only during or after construction. Correction is then costly or even impossible.

With VR based approaches, human factors engineers can explore the spaces and volumes of the plant early in the design process and verify that workers have adequate access to equipment that they need to interact with and enough space to perform what they need to do. Configurable digital manikins may be used to assess visibility, reachability, fatigue and other HFE concerns effectively and accurately. Two dimensional design information and plant process simulation data may be integrated in addition to 3-D geometric data, to create an operative and dynamic virtual plant. Also, some of the HFE design guidelines could be verified automatically.

6.9.2.2. *Disaster preparedness and response training*

VR based training provides trainees with a virtual but realistic environment where they can learn and rehearse the actions they might need to implement in the real world. It is a particularly good solution for disaster (e.g. fire, flood or earthquake) preparedness and response training, the conditions of which are hard to replicate in the real environment.

Trainers can configure and load different scenes as needed, for example, a flood with particular characteristics at a specific location in a virtual plant. Trainees can then take the full measure of the plant condition caused by the disaster and see, step by step, the effects of the emergency and evacuation procedures.

6.9.2.3. *Refuelling training*

Refuelling is often on the critical path of an outage. In the refuelling CAVE based training system shown in Fig. 14, a physical console is used to control a virtual operating vehicle and fuel grab. It provides a very high level of fidelity, not only in terms of haptics but also in terms of control logic. Different tasks or incidents can be planned and trained for. Except for the emergency manual mode that uses a virtual handwheel, all other modes are controlled with the physical console.

6.9.2.4. *Augmented reality based aids for mobile workers*

The AR device is fixed on the helmet to free a worker's hands. It associates objects of interest with relevant real time information, documents, operational data and trends, animated instructions, task procedures and records. It also provides the operator with audio and video communication with remote experts for help and support.

6.9.2.5. *Radiation visualization*

Radiation dose management is critical for operation and decommissioning, including planning, schedule optimization, inventory tracking and waste management. VR based training first helps field workers become familiar with the tasks to be performed. Then, AR maps radiation measurements and calculations in real time on the real environment, helping workers visualize radiation distribution during actual task implementation.

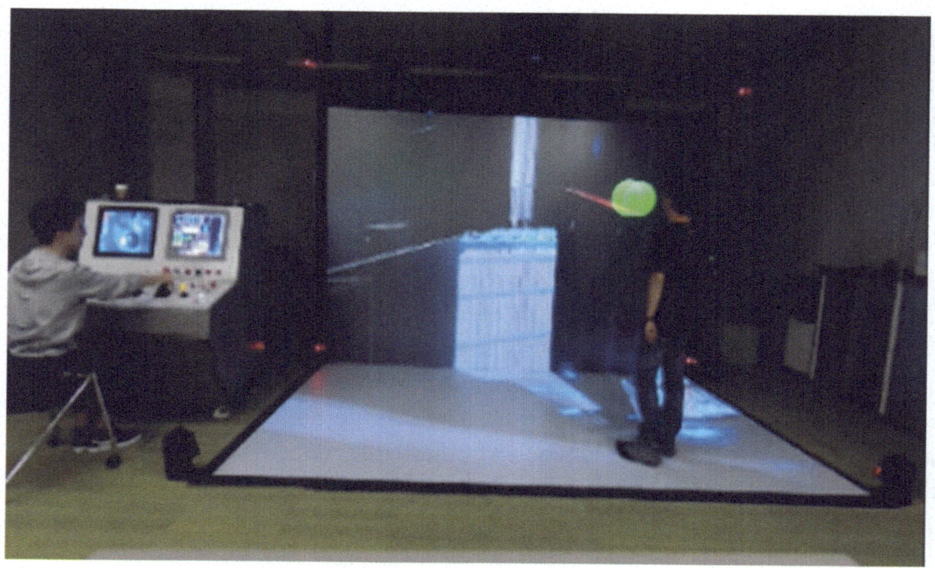

FIG. 14.   Refuelling semiphysical virtual reality based training system (courtesy of F. Song, China Nuclear Power Engineering Corporation, China).

6.9.3.   Maturity

VR is successfully applied by some in the nuclear industry, but for limited applications. AR is still new in the nuclear industry; there are few cases of applications and some of them are still in research and development. The nuclear industry could learn from the experience and successes of other industrial sectors, such as the gaming, real estate, healthcare, and automotive and aviation industries.

6.10. THREE DIMENSIONAL PRINTING

Three dimensional printing (also called additive manufacturing) is the construction of three dimensional objects from digital 3-D models. It may be applied to a wide variety of materials and can generate complex shapes that would be difficult to produce with more traditional means, with a high degree of accuracy and quality.

6.10.1. Benefits to plant performance

Experience shows that 3-D printing can:

— Streamline design processes.
— Increase design options for enhanced functionality and improved performance.
— Deliver qualified components in a highly regulated environment.
— Be a convenient and cost effective means for producing one of a kind or few of a kind objects (e.g. for prototyping or to support long term operation by the manufacturing of spare parts that can no longer be found on the market).

6.10.2. Example

One illustrative example of an application of 3-D printing within the nuclear industry is 3-D printed fuel assembly components. Fuel assembly components can be 3-D printed to increase their integrity or to produce new unconventional designs. Three dimensional printed stainless steel components are already in operation in nuclear power plants. For example, channel fasteners were loaded in a US reactor in 2021, and upper tie plate grids were loaded in a Swedish reactor in 2022 (see Fig. 15). Both components are located on top of the fuel assembly. Conventional processes would have required more manufacturing steps and human oversight compared to 3-D printing.

6.10.3. Maturity

Three dimensional printing is a relatively new technique, but it attracts a lot of attention in general industry, particularly for the production of small series of high quality objects with complex shapes.

6.11. CONSTRUCTION DATA INTEGRATION

Many different teams, disciplines, organizations and stakeholders are involved in the development of a nuclear power plant and of its systems. Their numerous interdependencies and challenging time schedules require rigorous and efficient technical coordination, not only for initial design and construction but also for the modifications, upgrades and ultimately decommissioning that inevitably occur during a plant's lifetime. The integration of the documents, plans, requirements, assumptions, models, simulation and analysis results, test plans and test results, as-built characteristics, laser scans and other types of data that these involved parties produce is an essential element of that coordination. The challenges are, on the one hand, the huge amount of data to be considered and, on the other hand, the large number and diversity of data sources.

A data integration platform (DIP) helps address these challenges with a structured data repository that collects, stores, links, compares and references data assets from all these sources to provide services, such as:

— Product life cycle management (i.e. the management of all data assets related to the design, construction, servicing and disposal of a product — here, the nuclear power plant and its systems);
— Data version and configuration management (e.g. the upkeep of all data assets and their organization into versions and configuration baselines);

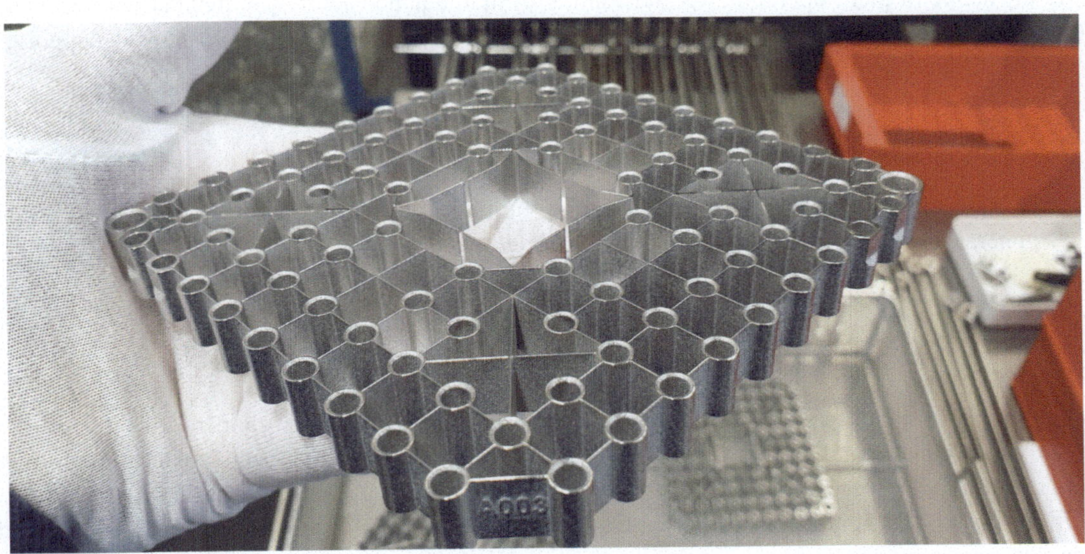

FIG. 15. *Three dimensional printed upper tie plate grid (courtesy of T. Salnikova, Framatome, Germany).*

— Change management (i.e. the scheduling and tracking of change requests and their links to actions and concerned plant items and data assets);
— Access control, so that authorized users can access data assets, but only according to their rights.
— Linkage between data pieces and plant items;
— Data integration by the establishment of links between related data pieces, explicitly connecting each piece with its 'context';
— Consistency checks between data pieces and their context;
— Visualization of data within or in relationship to their context;
— Reporting on and visualization of project status and key performance indicators and critical paths in time schedules.

6.11.1. Benefits to plant performance

For new build, modernization or outage projects, a DIP can significantly reduce lead time, costs and risks with:

— The availability of all the data in one place, with defined maturity level and consistency, avoiding time consuming searches or comparisons;
— Efficient coordination of the multiple teams, disciplines, organizations and stakeholders involved and reduced inconsistencies at their interfaces;
— Early detection of quality issues and impact of modifications, avoiding late detection of non-conformities and missing or obsolete data;
— Automated creation of documents, including data and reports, with the help of predefined templates.

6.11.2. Example

One illustrative example of an application of construction data integration within the nuclear industry is data integration of design and actual instrumentation (see Fig. 16).

As indicated by the arrows in Fig. 16, any addition or modification of DIP data will result in either a request to link the addition or modification to other database items or the automatic update of existing data. In terms of quality assurance, data quality checks can be defined either by comparing data from different sources or by verifying predefined engineering rules or requirements.

In the scope of plant operation, the scheduling of maintenance actions on plant structures, systems and components, as well as the sequence and dependence of assignments, can be visualized, supported by respective tasks and work order assignments that cut outage times.

6.11.3. Maturity

DIPs have demonstrated their value for new build and major modernization projects. Smaller projects generally need or use only selected DIP features.

6.12. COMMON INFORMATION MODELS

Common information models are standards specifying how the managed elements of a given industry sector are represented in a digital environment. Their objective is to facilitate and streamline information exchange, not only between software tools from different vendors but also between facility operators.

IEC 61970 [103] is a series of international standards based on an electric power transmission and distribution common information model named Common Information Model (CIM). It has been used by the electric power industry with great success for more than 20 years. However, as it does not cover

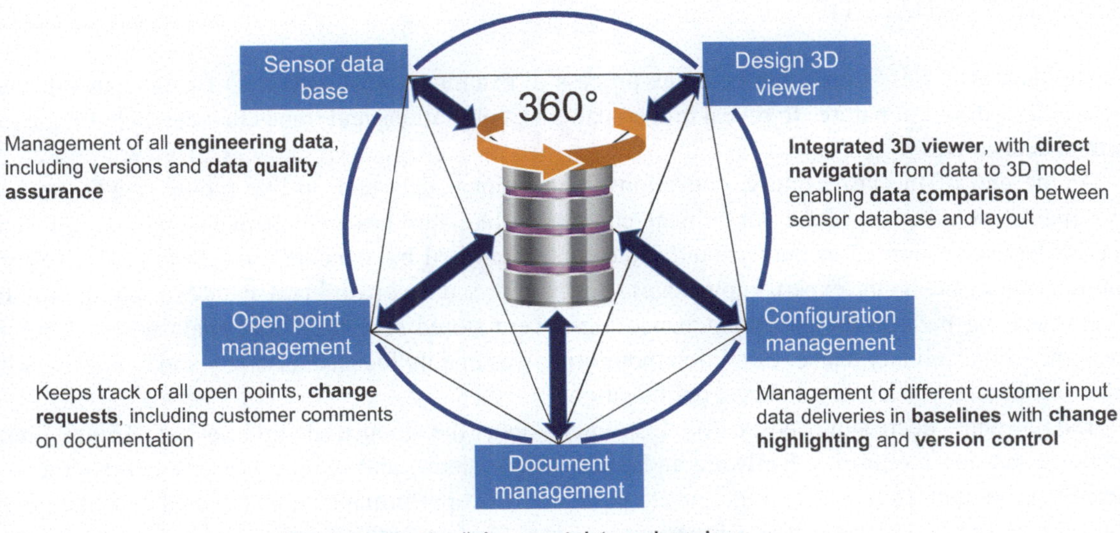

FIG. 16. Integration of design and actual instrumentation in a DIP (courtesy of F. Weser, Framatome, Germany).

power generation assets in sufficient detail, an extension named Energy Supply Common Information Model (ES-CIM) [104] has been developed, in particular to better support nuclear modernization.

6.12.1. Benefits

Improved information exchange generally improves not only the performance of individual facilities but also the overall performance of interconnected facilities and fleets.

6.12.2. Examples

The following are some illustrative examples of applications of common information models:

(a) The Common Grid Model Exchange Standard (CGMES): This CIM based data exchange is used by European transmission system operators to share the topology and state of their networks with each other and their cooperation organization, the European Network of Transmission System Operators for Electricity (ENTSO-E) [105].
(b) Data exchange between facilities: On the distribution side, a recent interoperability test confirmed the ability to exchange data between facility design, grid model management, geographic information system (GIS) and power flow applications in the distribution domain (see Ref. [106]).

6.12.3. Maturity

The maturity of the CIM for plant information exchange is evolving. Currently, interfaces are defined for water chemistry, heat transfer and maintenance order.

6.13. BIG DATA ANALYTICS

Big data analytics is the often complex process of examining vast amounts of data from different sources and of different natures to reveal hidden patterns, correlations and trends that can help in making informed decisions.

At the end of the last century, collection of operational data and on-line monitoring in nuclear power plants were mostly done for critical equipment like turbines and main coolant pumps. The collected data were limited to known fault indicators (identified by standards like Refs [107, 108] for spinning equipment, or by expert knowledge). Furthermore, data of different natures (condition data, process/operating parameters and maintenance logs) were stored separately and analysed by different experts and/or disciplinary teams. Condition monitoring was mainly focused on safety and not necessarily on plant availability, efficiency and cost optimization.

Today, with decreasing costs for instrumentation, the industrial Internet of Things, data collection/condition monitoring hardware and computing systems, and robotic process automation (see Annex II), more data from assets important to plant and fleet performance are collected and monitored on-line in addition to data from assets important to safety. Consequently, the objectives of data collection are no longer restricted to safety but are increasingly extended to security (including computer security) and performance.

Finally, as the numbers of instrumented assets, connected systems and collected data in fleets/plants continue to grow, the demand for digital solutions to efficiently gather, store, share, protect and transform huge volumes and varieties of data is rising. According to Forrester Research, "between 60% and 73% of all data within an enterprise goes unused for analytics" [109].

6.13.1. Benefits to plant performance

Big data analytics can help reduce operational costs (e.g. by optimizing performance, maintenance intervals, maintenance duration, spare parts management and component replacement). The performance benefits can be realized with no additional cost for instrumentation and data collection, because big data analytics leverage the vast amount of data readily collected as part of routine operation. Insights from big data analytics can inform where station investments should be made for asset upgrades according to value based considerations.

6.13.2. Examples

There are plenty of examples where big data analytics solutions have been successfully deployed, have improved the plant performance and also led to significant cost savings. Annex II describes an example of big data analytics technology in more detail, and the following are some other illustrative examples.

6.13.2.1. Baker Hughes predictive maintenance software using data analytics and MATLAB

One very well known example of big data analytics is the predictive maintenance solution deployed by Baker Hughes. The company operates several trucks at different sites simultaneously, with pumps injecting a mixture of water and sand at high pressure deep into drilled wells. These pumps are costly, accounting for about 10% of the total cost of a truck. If a pump fails, Baker Hughes has to immediately replace the truck to ensure continuous operation. Sending spare trucks to each site costs the company millions of dollars in revenue that those trucks could generate if they were in active use at another site. The inability to predict when valves and pumps will require maintenance underpins other costs. Overly frequent maintenance wastes effort and results in parts being replaced when they are still usable, while too infrequent maintenance risks damaging pumps beyond repair.

To address this challenge, Baker Hughes engineers wanted to develop a system that could determine when a pump was about to fail and needed maintenance. To do this, the team needed to process and analyse one terabyte of high resolution data. Using MATLAB, the team was able:

— To convert previously unreadable pump data into a usable format;
— To automate signal processing and data transformation steps;
— Ultimately, to apply machine and deep learning techniques in real time to predict the ideal time to perform maintenance.

According to MathWorks [110], several industrial companies, including Shell, Mondi, SNCF, Safran and Daimler, use MATLAB data analytics modules to analyse big data sets to detect events and abnormalities at power, process and chemical plants.

6.13.2.2. *Prediction of steam generator clogging*

As a complement to the estimation of steam generator clogging (see the example of Section 4.3.2.4), a supervised machine learning approach, run over data representing the operation of 20 units of the EDF fleet for a decade, identified 14 clogging influencing factors and provided a prediction model of clogging evolution. The model has been used to define new operation rules in order to prevent clogging as well as to improve the planning of steam generator maintenance activities.

6.13.2.3. *Optimization of flexible load operation*

Nuclear power plant flexible load operation causes frequent changes in power level, that in turn cause more stress and fatigue on some assets of the secondary circuit, such as pumps, valves, pipes and heat exchangers. To keep the safety, reliability and economic performance of the nuclear power plant at an optimal level, several CMSs were installed (see Fig. 17) to acquire detailed asset data (vibrations, temperature, electrical signals) and process data (e.g. coolant temperatures/pressures/mass flows, power levels and valves states). With big data analytics, the recorded data are now automatically transformed into diagnostics and insights to improve plant availability (see also Annex II and Annex III–2.2).

6.13.3. Maturity

Big data analytics has already been deployed in several industrial and nuclear plants as an off-line data analytics and decision support tool. The data import/delivery (or communication interfaces) and the data science methods used (e.g. signal/natural language processing, statistics, machine learning and deep learning) are mature and proven techniques.

However, big data analytics is not widely used today in the nuclear industry, despite very good results and successes from research and development, pilot projects and non-nuclear industrial plants. There are numerous reasons or barriers that are not favourable to a wide deployment:

— Poor instrumentation of some plant components and lack of data (quantity and quality of the existing data, limited number of sensors and recorded signals, missing raw data, missing data with component faults);
— Collected asset data not related to critical asset faults (no failure mode and symptoms analysis done according to Ref. [111]);
— Collected data not time stamped;

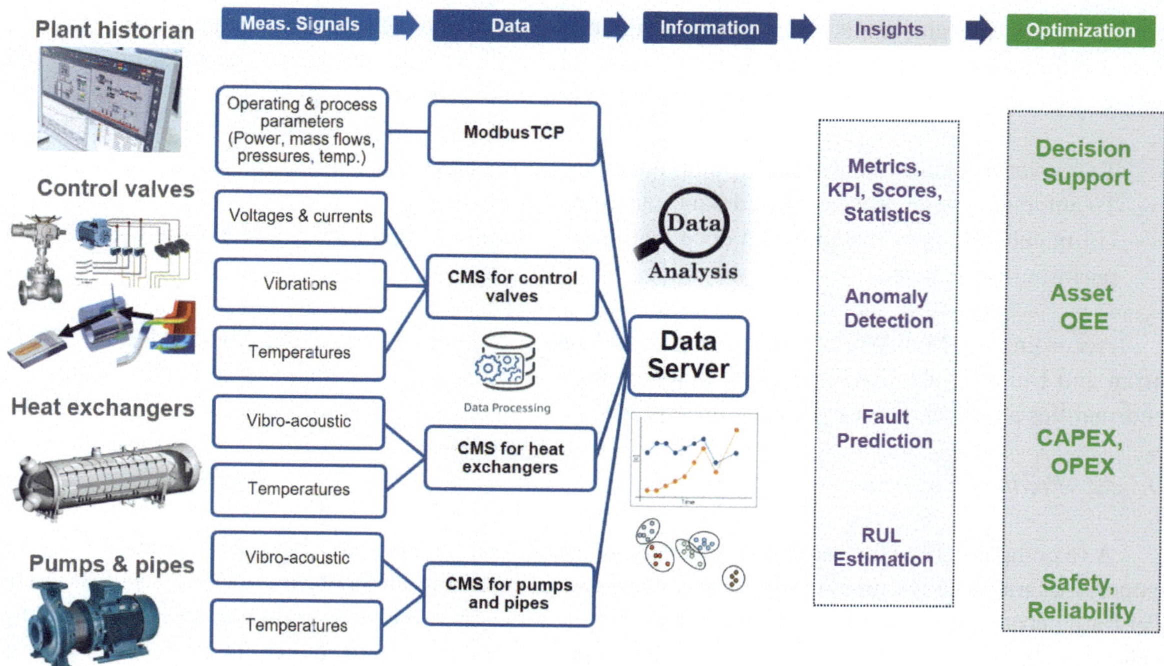

FIG. 17. Big data analytics application for flexible load operations. CAPEX — capital expenditure; CMS — core monitoring system; KPI — key performance indicator; OEE — overall equipment effectiveness; OPEX — operating expenditure; RUL — remaining useful life; TCP — transmission control protocol (courtesy of T. Salnikova, Framatome, Germany).

— Complex and difficult communication between existing data loggers, data sources like programmable logic controllers, supervisory control and data acquisition systems, computerized maintenance management systems and data historians (lack of standard communication protocols and computer security issues) from various vendors;
— Proprietary file formats, as is frequent in the nuclear industry (lack of open file formats);
— The cost of big data analytics platform solutions and questionable return on investment;
— The doubts of some plant operators regarding data confidentiality and data protection.

6.14. ARTIFICIAL INTELLIGENCE

The importance of AI in industrial applications is growing rapidly, thanks to progress in computing capacities, the availability of large amounts of data and improvement in AI algorithms. The nuclear industry has been applying AI techniques since at least the early 1980s. Reference [112] presents a view of AI systems as engineered systems that generate outputs, such as content, forecasts, recommendations or decisions, for a given set of human defined objectives.

There were initially two major approaches to AI:

— Symbolic AI encodes human knowledge with symbols and structures. It primarily uses logics or mathematical formulation to model reasoning processes, and its white box nature is a major advantage for nuclear applications. Expert systems are well known examples of symbolic AI that can match or exceed the performance of human experts on narrowly defined tasks, if given the appropriate domain knowledge.
— Subsymbolic AI lets the AI model learn from data gathered for this specific purpose. It relies on implicit knowledge hidden in data. Machine learning is a key area of subsymbolic AI, where a machine 'discovers' and 'learns' from data patterns. Machine learning can be subdivided into

supervised machine learning, unsupervised machine learning and reinforcement machine learning. Deep learning is a recent development of machine learning based on artificial neural networks. The concept is not new but has gained popularity and efficiency with extensive processing power and huge amounts of data.

Furthermore, an emerging hybrid approach can combine the benefits of symbolic AI and subsymbolic AI.

6.14.1. Benefits to plant performance

As demonstrated in the examples below, AI can contribute at each life cycle stage of a nuclear power plant (from design to operation and even decommissioning) to improve efficiency, reliability, safety and competitiveness. For example, it can help optimize design parameters or topologies; reduce the need for physical experimentation; decrease violation of safety rules during construction; and support advanced monitoring, detection, diagnostics and prognostics. It can reveal patterns in huge datasets and can then be used to accurately predict what will happen to structures, systems and components, facilitating preventive or even predictive maintenance. It can also forecast the occurrence of operational transients or abnormal operational conditions and alert plant operators.

6.14.2. Examples

The following are some illustrative examples of applications of AI in the nuclear power industry:

(a) AI for design and analysis of reactor processes: Thermohydraulic phenomena during transients are often very complex, and predicting turbulent behaviour is a challenging task that requires accurate and computationally efficient models. Coarse mesh 3-D computational fluid dynamics models are computationally efficient but not so accurate. Fine mesh computational fluid dynamics models are accurate but not so computationally efficient. Neural networks can supplement coarse mesh models by learning from data sets generated with fine mesh models, yielding a model that is both computationally efficient and accurate.

(b) AI for construction: When pouring large volumes of concrete, many phenomena (e.g. inhomogeneities in concrete performance, equipment problems or cement hydration heat) can lead to undesirable effects (e.g. cracks). Traditionally, identifying these phenomena relies on expert judgement based on many parameters, which is time consuming and may result in suspension of pouring operations. Supervised machine learning has been experimented with to automatically analyse concrete pouring data and give real time advice, saving time and improving concrete quality.

(c) AI to assist operation: Currently, nuclear power plant operation relies mostly on traditional systems (i.e. non-AI based) to detect abnormal situations, but warnings sometimes come too late to take actions that could prevent loss of performance. As algorithms have advanced and computing power has increased, some plants use AI for efficient operation, such as advanced and improved diagnostics and prognostics and high levels of automated operations. Supervised, unsupervised and semisupervised machine learning algorithms have been developed to detect subtle anomalies and small deviations in nuclear power plants. For example:

   (i) Autoencoders, unsupervised deep learning algorithms, have been trained to detect hidden patterns in a huge volume of operation data and then deployed to predict anomalies occurring in plant processes.

   (ii) Reinforcement learning is explored to perform automatic operation for heat-up mode, which is typically mostly manual (see Ref. [113]).

   (iii) In Annex V, AI is extensively explored for analysing time series data, to implement fault prediction (using deep learning classification models) and RUL/value prediction and/or prognostics (using deep learning regression and time series forecasting models).

(d)   AI for computer security: With the growing use of digital systems in the nuclear industry, computer security has become an increasingly important concern. Prevention, detection and reaction are key elements of a computer security programme. AI has been investigated for the detection of malicious attacks in nuclear power plants (see also Section 4.1.2). For detection, various AI methods can be applied, but because data related to computer security events are limited, unsupervised learning methods (which do not require two classes of tagged data, normal and abnormal) are generally the most appropriate.

6.14.3. Maturity

All these examples indicate that AI has matured enough to find extensive applications, as testified by Gartner in Ref. [114]. Another indicator is the creation in 2017 of the joint ISO/IEC technical committee JTC 1/SC 42, 'Artificial intelligence'. Additionally, IEC subcommittee SC 45A, 'Instrumentation, control and electrical power systems of nuclear facilities', is developing standards on the application of AI in the nuclear industry, the first of which is Ref. [115]. The IAEA published Ref. [116] in 1994 and organized several meetings in 2020 and 2021 to promote the adoption and use of AI.

However, AI is not without weaknesses. The complex and non-deterministic nature of many AI systems severely hinders their licensing. Thus, although some progress is being made in the testing (see Ref. [117]) and explainability of AI systems, more work is still necessary before AI can replace or augment a safety related or risk-significant system.

Reference [118] provides an overview of the applications and capabilities of AI applications deployed in nuclear power plants.

7. PRACTICAL CHALLENGES TO IMPLEMENTATION

Although digital I&C systems and other advanced digital technologies offer significant opportunities to enhance nuclear power plant performance, their implementation is not without challenges. As discussed in this section, these challenges include ensuring that innovation can progress within regulatory frameworks designed for previous generations of technologies, as well as managing organizational change in an industry traditionally focused on proven methods and incremental improvements. Long term considerations are critical to sustaining benefits over the plant life cycle, such as managing tool and model lifetimes and maintaining proficiency with rapidly evolving tools, methods and data. In addition, the increasing reliance on digital systems heightens the importance of robust computer security measures to protect safety, security and sensitive information against cyber threats.

7.1. INNOVATION WITHIN REGULATION

Innovation is one of the major impetuses for achieving long term economic growth, attaining competitiveness and providing solutions to many challenges in the nuclear and other industries. It has been a driving force for the success of the nuclear industry in the world and will remain critical for its sustainable future. In recent years, some Member States have launched national programmes to develop innovative technologies to further enhance safety and optimize performance in the nuclear industry.

However, besides its potential benefits for boosting safety, reliability and operational performance, innovation could also bring new and sometimes unknown risks to nuclear power plants. Regulation by independent governmental authorities is indispensable to ensure nuclear safety and security. It has been successfully implemented in the nuclear power industry from the industry's beginnings, but some innovations may pose challenges to existing regulations, which were created for previous generations of technologies and practices.

Although safety and security assurance is the primary objective of all nuclear regulatory authorities, the industry needs predictable but also adaptable regulatory frameworks that can adopt innovations beneficial to safety, security and economic benefits for existing and new nuclear power plants.

The future deployment of new and advanced reactors is highly expected to use innovative technologies and will benefit from regulatory frameworks that would be appropriately established to independently analyse and evaluate the safety implications of innovative technologies.

7.2. CHANGE MANAGEMENT

With its strong focus on safety and security, the nuclear industry tends to rely on well proven methods, equipment and solutions. Also, due to the high complexity of a nuclear power plant, plant designers tend to prefer evolutionary approaches, where a new plant design is obtained by introducing limited, well focused changes to an existing, proven-in-use design. With such approaches, designers tend to reuse the documents and methods of the original design. Naturally, there are innovations here and there, but generally not as the result of a systematic process.

However, as economic pressure from other sources of energy grows, the nuclear industry now has an urgent need to innovate at a much larger scale. This can be seen with the growing number of SMR projects, many of which feature innovations that were deemed unrealistic a few years ago. Also, other industries have opened the way and have pioneered the use of innovative approaches, methods, technologies and products, often with great success. In doing so, they have been able to reduce design schedules, streamline manufacturing and operation and offer innovative services and products. Advanced digital technologies, in particular, have a very strong impact. The nuclear industry now stands to benefit from following this example.

7.3. MANAGEMENT OF TOOL AND MODEL LIFETIMES

Effective management of tool and model lifetimes is essential to ensure continuity of support for design, construction and operational activities throughout the plant life cycle.

7.3.1. Tools

The services provided by digital tools for design and construction are often needed throughout the rest of the lifetime of a plant, plant systems or I&C systems in order to support operation, maintenance, outages, modifications and, ultimately, decommissioning. That constraint may need to be considered when selecting tools. However, that does not necessarily mean that a tool cannot evolve or be replaced. Indeed, it may sometimes be preferable to use a new, more powerful and better supported version, or different, more advanced and more powerful tools, provided they maintain upward compatibility and computer security, and provided that the tools supporting safety or computer security justifications are qualifiable. Upward compatibility is generally necessary to ensure that the new version or the new tool can use existing models and data.

In the management of tools' lifetimes, it is also necessary to include the hardware and software infrastructures necessary for their operation. Software infrastructures are sometimes extensive, with operating systems and general purpose frameworks.

7.3.2. Models

Models generally require significant development and validation efforts. With appropriate systems engineering planning, some models developed to support design can also be used in support of construction, operation and maintenance (see Sections 3.6–3.11, 4 and 6.9), provided that they are also maintained, ensuring that:

— They have been verified to be correct with respect to what they represent and adequate for the activities for which they are used.
— They are kept up to date with what they represent.
— Their data formats are updated when new tools or new tool versions using new data formats replace the original ones.
— They are upgraded as necessary as new uses and/or new services are put in place.

7.4. PROFICIENCY WITH TOOLS, METHODS AND DATA

Maintaining proficiency with rapidly evolving digital tools, methods and data is essential to ensure continued benefits from performance enhancements and to support safe, efficient plant operation.

7.4.1. Tools and methods

Digital technologies (for I&C and advanced design and/or operational aids) evolve very rapidly compared to the lifetime of a nuclear power plant. This raises the issue of maintaining or acquiring the necessary competences on the digital tools supporting operation and possibly decommissioning, considering that, often, they are not just software tools, but may also include hardware (e.g. mobile communication devices) and rely on a supporting infrastructure (e.g. an operating system and/or a communication network).

Continued benefit from the plant performance enhancements enabled by these tools depends on the application of training (see Ref. [119]) and knowledge management (see Refs [120, 121]) that address not only the tools themselves but also their underlying methods and their supporting infrastructures.

7.4.2. Data

With the widespread use of digital technologies, data have become an essential asset, to the point that they are at the heart of the business model of hugely successful firms such as the 'Big Five' technology companies (Google, Amazon, Meta, Apple and Microsoft). This is also becoming true for nuclear power plants, where vast amounts of data are collected from a multitude of design, construction and operation sources, and possibly from many plants and units in the case of a fleet. Training and knowledge management also need to be applied to fully benefit from such data, which need to be captured, stored, described and organized, understood, preserved in the long term, protected and possibly shared with authorized stakeholders.

7.5. COMPUTER SECURITY

Digital technologies are increasingly being incorporated into and integrated with I&C systems. New nuclear facilities and modern nuclear facility designs are using highly integrated digital I&C systems capable of handling large quantities of process data, reducing the need for human interaction and intervention. Computer based systems (i.e. systems that make use of, depend on or are supported by digital technologies), play an ever expanding role in the performance management of sensitive information, nuclear safety, nuclear security and nuclear material accounting and control at nuclear power plants.

Computer based systems, including advanced technologies, are used to perform operational and administrative functions at nuclear facilities and are increasingly used to support a wide range of diverse activities including design, maintenance, procurement and performance optimization. Cyber-attacks on I&C systems and other computer based systems may jeopardize the safety and security of nuclear facilities. They may contribute to sabotage or aid in the unauthorized removal of nuclear material. The effects of cyber-attacks on systems related to safety may result in a wide range of consequences, such as a temporary loss of process control or unacceptable radiological consequences. Therefore, computer security is needed for the prevention and detection of cyber-attacks, as well as response to and recovery of computer based systems from cyber-attacks.

Cyber-attacks may modify the configuration, programming or data within a computer based system and may deceive the users of the system by modifying data that are presented to users, leading to incorrect actions being undertaken following operating procedures. For instance:

— A cyber-attack against an I&C system could modify the configuration, logic or set points of an I&C system and prevent the actuation of necessary actions. Such attacks may put the plant in a state that has not been considered in any safety analysis report. The potential consequences of a compromise on I&C system functions are, arranged in the order of worst to best case scenario, the following:
  - The performance of the function is indeterminate. This means that the function might be altered in any manner without the initial compromise being detected.
  - The performance of the function changes in unexpected ways (and other actions can be performed), but these anomalies are observable to the operator.
  - The performance of the function fails.
  - The performance of the function is as expected, meaning the compromise does not adversely affect the function (i.e. the system is fault tolerant).
— A cyber-attack against a system used for performance based maintenance of plant equipment may be modified to present misleading information about the state of some plant equipment. As a result, the maintenance organization may fail to perform the necessary maintenance.
— A cyber-attack targeting sensitive information about the design of the physical protection system could be used by an adversary to facilitate attacks aimed at the theft of nuclear material or sabotage.

Guidance on computer security for nuclear facilities can be found in Ref. [13], which describes how computer security risk management is incorporated within the management system throughout the lifetime of the facility. Reference [13] describes a systematic approach for identifying facility functions that require protection (i.e. those relevant to the safe and secure operation of the facility) and provides guidance on determining the computer security requirements for the systems that implement or support these functions. It provides guidance specific to nuclear facilities on implementing a computer security programme, on defining and implementing defensive computer security architecture that provides for defence in depth, and on applying computer security measures in a graded approach.

Reference [14] provides specific computer security guidance for the protection of I&C systems at nuclear facilities throughout their life cycle against malicious acts that could prevent such systems from performing their safety and security related functions. It also addresses the application of computer security measures to I&C systems. This includes guidance on a risk informed approach to computer security for I&C systems and on a graded approach and defence in depth, including the application of computer security measures (technical control, physical control or administrative control measures) to address vulnerabilities in these systems.

Reference [122] addresses reducing risk within the nuclear supply chain. Due to the complexity of the supply networks for digital technologies used at nuclear facilities, there are numerous possibilities for an adversary to compromise a service or device. Reference [122] describes a supply chain attack surface and identifies mitigation techniques to reduce it throughout the life cycle of the supply of the product or service.

8. CONCLUSION

Digital I&C is nowadays common practice in new builds, and many operating plants have replaced all or part of their conventional I&C with digital I&C. Non-I&C advanced digital technologies are also used in many plants and projects, as illustrated by the many examples presented in the preceding sections and in the annexes. However, these technologies are generally not exploited to their full potential, and the nuclear industry still has significant room for further progress.

To fully realize the benefits of advanced digital technologies, it is best to embed their use within a systems engineering framework (see Ref. [22]), so that they can also be of benefit to teams, engineering disciplines, organizations and stakeholder coordination.

REFERENCES

[1] INTERNATIONAL ATOMIC ENERGY AGENCY, Safety of Nuclear Power Plants: Design, IAEA Safety Standards Series No. SSR-2/1 (Rev. 1), IAEA, Vienna (2016).

[2] INTERNATIONAL ATOMIC ENERGY AGENCY, Design of Instrumentation and Control Systems for Nuclear Power Plants, IAEA Safety Standards Series No. SSG-39, IAEA, Vienna (2016).

[3] INTERNATIONAL ATOMIC ENERGY AGENCY, Effective Corrective Actions to Enhance Operational Safety of Nuclear Installations, IAEA TECDOC-1458, IAEA, Vienna (2005).

[4] ELECTRIC POWER RESEARCH INSTITUTE, The Economics of Nuclear Plant Modernization in U.S. Markets, Rep. 3002014737, EPRI, Palo Alto, CA (2019).

[5] INTERNATIONAL ATOMIC ENERGY AGENCY, Dependability Assessment of Software for Safety Instrumentation and Control Systems at Nuclear Power Plants, IAEA Nuclear Energy Series No. NP-T-3.27, IAEA, Vienna (2018).

[6] INTERNATIONAL ELECTROTECHNICAL COMMISSION, Nuclear Power Plants — Instrumentation and Control Important to Safety — General Requirements for Systems, IEC 61513:2011, IEC, Geneva (2011).

[7] INTERNATIONAL ELECTROTECHNICAL COMMISSION, Nuclear Power Plants — Instrumentation and Control Systems Important to Safety — Software Aspects for Computer-based Systems Performing Category A Functions, IEC 60880:2006, IEC, Geneva (2006).

[8] INTERNATIONAL ELECTROTECHNICAL COMMISSION, Nuclear Power Plants — Instrumentation and Control Systems Important to Safety — Software Aspects for Computer-based Systems Performing Category B or C Functions, IEC 62138:2018, IEC, Geneva (2018).

[9] INTERNATIONAL ELECTROTECHNICAL COMMISSION, Nuclear Power Plants — Instrumentation and Control Important to Safety — Hardware Requirements, IEC 60987:2021, IEC, Geneva (2021).

[10] INTERNATIONAL ELECTROTECHNICAL COMMISSION, Nuclear Power Plants — Instrumentation and Control Important to Safety — Development of HDL-programmed Integrated Circuits for Systems Performing Category A Functions, IEC 62566:2012, IEC, Geneva (2012).

[11] INTERNATIONAL ELECTROTECHNICAL COMMISSION, Nuclear Power Plants — Instrumentation and Control Systems Important to Safety — Development of HDL-programmed Integrated Circuits — Part 2: HDL-programmed Integrated Circuits for Systems Performing Category B or C Functions, IEC 62566-2:2020, IEC, Geneva (2020).

[12] INTERNATIONAL ATOMIC ENERGY AGENCY, Computer Security for Nuclear Security, IAEA Nuclear Security Series No. 42-G, IAEA, Vienna (2021).

[13] INTERNATIONAL ATOMIC ENERGY AGENCY, Computer Security Techniques for Nuclear Facilities, IAEA Nuclear Security Series No. 17-T (Rev. 1), IAEA, Vienna (2021).

[14] INTERNATIONAL ATOMIC ENERGY AGENCY, Computer Security of Instrumentation and Control Systems at Nuclear Facilities, IAEA Nuclear Security Series No. 33-T, IAEA, Vienna (2018).

[15] INTERNATIONAL ATOMIC ENERGY AGENCY, Nuclear Power Plant Personnel Training and its Evaluation: A Guidebook, Technical Reports Series No. 380, IAEA, Vienna (1996).

[16]   INTERNATIONAL ATOMIC ENERGY AGENCY, Human Factors Engineering Aspects of Instrumentation and Control System Design, IAEA Nuclear Energy Series No. NR-T-2.12, IAEA, Vienna (2021).

[17]   INTERNATIONAL ATOMIC ENERGY AGENCY, Handbook on Ageing Management for Nuclear Power Plants, IAEA Nuclear Energy Series No. NP-T-3.24, IAEA, Vienna (2017).

[18]   HU, C., ZHOU, Z., ZHANG, J., SI, X., A survey on life prediction of equipment, Chin. J. Aeronaut. **28** 1 (2015) 25–33,
https://doi.org/10.1016/j.cja.2014.12.020

[19]   KALAYCI, C.B., KARAGOZ, S., KARAKAS, Ö., Soft computing methods for fatigue life estimation: A review of the current state and future trends, Fatigue Fract. Eng. Mater. Struct. **43** (2020) 2763–2785,
https://doi.org/10.1111/ffe.13343

[20]   TIAN, Z., An artificial neural network method for remaining useful life prediction of equipment subject to condition monitoring, J. Intell. Manuf. **23** (2009) 227–237,
https://doi.org/10.1007/s10845-009-0356-9

[21]   INTERNATIONAL ATOMIC ENERGY AGENCY, Management of Ageing and Obsolescence of Instrumentation and Control Systems and Equipment in Nuclear Power Plants and Related Facilities Through Modernization, IAEA Nuclear Energy Series No. NR-T-3.34, IAEA, Vienna (2022).

[22]   INTERNATIONAL ATOMIC ENERGY AGENCY, Introduction to Systems Engineering for the Instrumentation and Control of Nuclear Facilities, IAEA Nuclear Energy Series No. NR-T-2.14, IAEA, Vienna (2022).

[23]   INTERNATIONAL ATOMIC ENERGY AGENCY, Non-baseload Operation in Nuclear Power Plants: Load Following and Frequency Control Modes of Flexible Operation, IAEA Nuclear Energy Series No. NP-T-3.23, IAEA, Vienna (2018).

[24]   INTERNATIONAL ATOMIC ENERGY AGENCY, Progress on Pellet–Cladding Interaction and Stress Corrosion Cracking, IAEA-TECDOC-1960, IAEA, Vienna (2021).

[25]   KOSOWSKI, K., DIERCKS, F., Quo vadis, grid stability? Challenges increase as generation portfolio changes, ATW-Int. J. Nucl. Power **66** 2 (2021),
https://kernd.de/wp-content/uploads/2023/08/Article-atw-2021-2-Quo-vadis-Grid-Stability-Kosowski-Diercks.pdf

[26]   INTERNATIONAL ATOMIC ENERGY AGENCY, Non-electric Applications, IAEA, Vienna (2022),
https://www.iaea.org/topics/non-electric-applications/

[27]   INTERNATIONAL ATOMIC ENERGY AGENCY, Opportunities for Cogeneration with Nuclear Energy, IAEA Nuclear Energy Series No. NP-T-4.1, IAEA, Vienna (2017).

[28]   INTERNATIONAL ATOMIC ENERGY AGENCY, Guidance on Nuclear Energy Cogeneration, IAEA Nuclear Energy Series No. NP-T-1.17, IAEA, Vienna (2019).

[29]   KIM, K.K., LEE, W., CHOI, S., KIM, H.R., HA, J., SMART: The first licensed advanced integral reactor, J. Energy Power Eng. **8** (2014) 94–102.

[30]   WORLD NUCLEAR ASSOCIATION, Desalination, WNA, London (Mar. 2020),
https://world-nuclear.org/information-library/non-power-nuclear-applications/industry/nuclear-desalination.aspx

[31]   BRUCE POWER, Isotope Production System begins commercial production of cancer-fighting lutetium-177, Tiverton, ON (24 Oct. 2022),
https://www.brucepower.com/2022/10/24/isotope-production-system-begins-commercial-production-of-cancer-fighting-lutetium-177/

[32]   MOROKHOVSKYI, V., Reactor core control based on artificial intelligence, ATW-Int. J. Nucl. Power **65** 6/7 (2020) 350–352.

[33]   MOROKHOVSKYI, V., "Reactor core control based on artificial intelligence", paper presented at Int. Conf. on Nuclear Engineering & ASME Power Conf. (ICONE-POWER 2020), Anaheim, CA, 2020.

[34]   DeGROOT, M.H., A conversation with George Box, Stat. Sci. **2** 3 (1987) 239–258,
https://doi.org/10.1214/ss/1177013223

[35]   INTERNATIONAL ORGANIZATION FOR STANDARDIZATION, INTERNATIONAL ELECTROTECHNICAL COMMISSION, INSTITUTE OF ELECTRICAL AND ELECTRONICS ENGINEERS, Systems and Software Engineering — System Life Cycle Processes, ISO/IEC/IEEE 15288:2015, ISO/IEC/IEEE, Geneva (2015).

[36]   INTERNATIONAL ORGANIZATION FOR STANDARDIZATION, INTERNATIONAL ELECTROTECHNICAL COMMISSION, INSTITUTE OF ELECTRICAL AND ELECTRONICS ENGINEERS, Systems and Software Engineering — Methods and Tools for Model-based Systems and Software Engineering, ISO/IEC/IEEE 24641:2023, ISO/IEC/IEEE, Geneva (2023).

[37] MODELICA ASSOCIATION, Functional Mock-up Interface (FMI) Specification, Version 3.0, Modelica, Munich (May 2022),
https://fmi-standard.org/docs/3.0/

[38] HOEFER, A., FELDMANN, H., GLAUBRECHT, S., "Burnup credit criticality safety analysis for high density spent fuel storage using a unified Bayesian validation methodology", Proc. Nuclear Criticality Safety Division Topical Mtg (NCSD 2022), Anaheim, CA, 12–16 Jun. 2022, American Nuclear Society, La Grange Park, IL (2022) 148–157,
https://www.ans.org/pubs/proceedings/article-51932/

[39] EUROPEAN NUCLEAR SOCIETY, Top Fuel 2018 (Conf. Proc., full papers), Prague, Czech Republic, 30 Sep.–4 Oct. 2018, ENS, Brussels (2018),
https://old.euronuclear.org/events/topfuel/topfuel2018/proceedings.htm

[40] INTERNATIONAL ATOMIC ENERGY AGENCY, Phenomenology, Simulation and Modelling of Accidents in Spent Fuel Pools, IAEA-TECDOC-1949, IAEA, Vienna (2021).

[41] MODELICA ASSOCIATION, Modelica - A Unified Object-Oriented Language for Systems Modeling: Language Specification, Version 3.5, Modelica, Munich (Feb. 2021),
https://modelica.org/documents/MLS.pdf

[42] INTERNATIONAL ATOMIC ENERGY AGENCY, Power Uprate in Nuclear Power Plants: Guidelines and Experience, IAEA Nuclear Energy Series No. NP-T-3.9, IAEA, Vienna (2011).

[43] LANGENSTEIN, M., et al., "Finding megawatts in nuclear power plants with process data reconciliation", Proc. 12th Int. Conf. on Nuclear Engineering (ICONE12), Arlington, VA, 25–29 Apr. 2004, Vol. 2, ASME, New York (2004) 43–52,
https://doi.org/10.1115/ICONE12-49152

[44] INTERNATIONAL ATOMIC ENERGY AGENCY, Living Probabilistic Safety Assessment, IAEA-TECDOC-1106, IAEA, Vienna (1999).

[45] SÄTEILYTURVAKESKUS, Probabilistic Risk Assessment and Risk Management of a Nuclear Power Plant, Regulatory Guides on Nuclear Safety, STUK Guide YVL A.7, STUK, Vantaa, Finland (2019).

[46] WENSAUER, A., KNOLL, A., MARX, V., DEUTLE, D., SCHLIECK-WEBER, M., "Full-core statistical approach in LB-LOCA analysis – application to a core with non-nominal water gaps", paper presented at Annual Mtg of German Nuclear Society, Berlin, 2019,
https://www.researchgate.net/publication/332979124_Full-Core_Statistical_Approach_in_LB-LOCA_Analysis_-_Application_to_a_Core_with_Non-Nominal_Water_Gaps

[47] KNOLL, A., WENSAUER, A., DEUBLE, D., "Application of a full-core statistical approach in LB-LOCA analysis", Proc. 249th, ANS Best Estimate Plus Uncertainty Int. Conf. (BEPU 2018), Lucca, Italy, 13–19 May 2018.

[48] ELECTRIC POWER RESEARCH INSTITUTE, NextGen RP – Specifications for Autonomous Contamination Survey Robots: Functional Specifications Document, Rep. 3002023977, EPRI, Palo Alto, CA (2022).

[49] ELECTRIC POWER RESEARCH INSTITUTE, Autonomous Indoor Radiation Survey and Inspection Drone Demonstration at Peach Bottom, Rep. 3002018409, EPRI, Palo Alto, CA (2020).

[50] TROCHIANI, S., BUSS, O., HAUSSLER, S., LUDWIK, O., HOEFER, A., "AREVA computer codes for radiological consequence analysis", Presentation at the IAEA, Vienna, 20–24 Apr. 2015, AREVA, Offenbach, Germany,
https://www-pub.iaea.org/iaeameetings/IEM9p/Session2/6Torchiani.pdf

[51] BLANCHON, J.-C., HEMERY, G., Simulators as a design tool, Nucl. Eng. Int. (Nov. 2021) 30–31,
https://content.yudu.com/web/442ay/0A444i1/NEI1121-Pros/html/index.html?page=30

[52] GAUTER, K., FUCHS, T., "Simulation and design strategy for the development of advanced fuel pool cooling systems", Phenomenology, Simulation and Modelling of Accidents in Spent Fuel Pools, IAEA-TECDOC-1949, IAEA, Vienna (2021) 41–55.

[53] INTERNATIONAL ATOMIC ENERGY AGENCY, Advanced Surveillance, Diagnostic and Prognostic Techniques in Monitoring Structures, Systems, and Components in Nuclear Power Plants, IAEA Nuclear Energy Series No. NP-T-3.14, IAEA, Vienna (2013).

[54] INTERNATIONAL ATOMIC ENERGY AGENCY, Application of Wireless Technologies in Nuclear Power Plant Instrumentation and Control Systems, IAEA Nuclear Energy Series No. NR-T-3.29, IAEA, Vienna (2020).

[55] INTERNATIONAL ATOMIC ENERGY AGENCY, On-line Monitoring for Improving Performance of Nuclear Power Plants Part 1: Instrument Channel Monitoring, IAEA Nuclear Energy Series No. NP-T-1.1, IAEA, Vienna (2008).

[56] INTERNATIONAL ATOMIC ENERGY AGENCY, On-line Monitoring for Improving Performance of Nuclear Power Plants Part 2: Process and Component Condition Monitoring and Diagnostics, IAEA Nuclear Energy Series No. NP-T-1.2, IAEA, Vienna (2008).

[57] INTERNATIONAL ATOMIC ENERGY AGENCY, Technical Challenges in the Application and Licensing of Digital Instrumentation and Control Systems in Nuclear Power Plants, IAEA Nuclear Energy Series No. NP-T-1.13, IAEA, Vienna (2015).

[58] INTERNATIONAL ATOMIC ENERGY AGENCY, Condition Monitoring and Incipient Failure Detection of Rotating Equipment in Research Reactors, IAEA-TECDOC-1920, IAEA, Vienna (2020).

[59] INTERNATIONAL ATOMIC ENERGY AGENCY, Implementation Strategies and Tools for Condition Based Maintenance at Nuclear Power Plants, IAEA-TECDOC-1551, IAEA, Vienna (2007).

[60] INTERNATIONAL ATOMIC ENERGY AGENCY, Benchmark Analysis for Condition Monitoring Test Techniques of Aged Low Voltage Cables in Nuclear Power Plants, IAEA-TECDOC-1825, IAEA, Vienna (2017).

[61] HASHEMIAN, H.M., SHUMAKER, B.D., MORTON, G.W., Online Monitoring Technology to Extend Calibration Intervals of Nuclear Plant Pressure Transmitters, AMS Topical Report AMS-TR-0720R2-A, NRC ADAMS Accession No. ML21235A493, Analysis and Measurement Services Corporation, Knoxville, TN (2021).

[62] GOCHT, U., et al., "Data validation use for the surveillance and condition-oriented maintenance of heat exchangers in NPPs" (Annual Mtg on Nuclear Technology 2011, Documentation, Jahrestagung Kerntechnik 2011, Berlin, 17–19 May 2011), Kerntechnische Gesellschaft e.V., Bonn (2011).

[63] JUNG, J.C., SEONG, P.H., Error analysis in improved motor control center method for stem thrust estimation of motor-operated valves in nuclear power plants, IEEE Trans. Nucl. Sci. **50** 3 (2003) 735–740,
https://www.doi.org/10.1109/TNS.2003.812458

[64] DEMAZIÈRE, C., GLÖCKLER, O., "On-line determination of the prompt fraction of in-core neutron detectors in CANDU reactors", Proc. Int. Conf. on Physics of Reactors (PHYSOR 2004), San Francisco, CA, 2004, American Nuclear Society, La Grange Park, IL (2004).

[65] ELECTRIC POWER RESEARCH INSTITUTE, Mechanical Seal Failure Prevention and Condition Monitoring, Rep. 3002017901, EPRI, Palo Alto, CA (2020).

[66] EL BOUZIDI, S., RECTOR, T., "Using artificial intelligence to monitor pump seal degradation", Proc. Canadian Nuclear Society Symposium for Artificial Intelligence, Machine Learning, and Other Innovative Technologies (SAIMIN-2020), Toronto, 2020, Canadian Nuclear Society, Toronto (2020).

[67] VILIM, R., NGUYEN, T., PONCIROLI, R., "Explainable and trustworthy diagnostics achievable through process-based automated reasoning", Proc. 12th Nuclear Plant Instrumentation, Control and Human-Machine Interface Technologies (NPIC&HMIT 2021), Virtual Mtg, 14–17 Jun. 2021, American Nuclear Society, La Grange Park, IL (2021) 457–465.

[68] TAYADE, A., PATIL, S., PHALLE, V., KAZI, F., POWAR, S., Remaining useful life (RUL) prediction of bearing by using regression model and principal component analysis (PCA) technique, Vibroeng. Procedia **23** (2019) 30–36,
https://doi.org/10.21595/vp.2019.20617

[69] KIGER, C. J., et al., Implementation of new cable condition-monitoring technology at Oyster Creek Nuclear Generating Station, Nucl. Technol. **200** 2 (2017) 93–105,
https://doi.org/10.1080/00295450.2017.1360716

[70] ELECTRIC POWER RESEARCH INSTITUTE, Southern and EPRI Conduct Pilot Project for Fleetwide Monitoring and Diagnostic Center, Rep. 1020935, EPRI, Palo Alto, CA (2012),
https://www.epri.com/research/products/000000000001020935

[71] GERTMAN, D.I., LE BLANC, K., BORING, R.L., "Review of computerized procedure guidelines for nuclear power plant control rooms", Proc. Human Factors and Ergonomics Society 55th Annual Mtg, Las Vegas, NV, 2011, Human Factors and Ergonomics Society, Santa Monica, CA (2011) 1476–1480.

[72] SCHÄFLEIN H., KALEVA, H., ROTH-SEEFIRID, H., "Emergency operating procedures for Olkiluoto 3 NPP", Proc. Annual Congress for Nuclear Technology, Berlin, 4–6 May 2010, Kerntechnische Gesellschaft e.V., Bonn (2010).

[73] INTERNATIONAL ELECTROTECHNICAL COMMISSION, Nuclear Power Plants — Control Rooms — Computer-based Procedures, IEC 62646:2016, IEC, Geneva (2016).

[74] ELECTRIC POWER RESEARCH INSTITUTE, Unmanned Aircraft System (UAS) User's Guide for Nuclear Power Plants: Implementation Guidance, Technologies and Applications, and Cost Savings Opportunities, Rep. 3002020913, EPRI, Palo Alto, CA (2021).

[75] ELECTRIC POWER RESEARCH INSTITUTE, Automated Analysis of Remote Visual Inspection of Containment Buildings, Rep. 3002018419, EPRI, Palo Alto, CA (2020).

[76] RAY, A., A microcomputer-based fault-tolerant control system for industrial applications, IEEE Trans. Ind. Appl. **IA-21** 5 (1985),
https://doi.org/10.1109/TIA.1985.349555

[77] GARCIA, H.E., RAY, A., A reconfigurable hybrid system and its application to power plant control, IEEE Trans. Control Syst. Technol. **3** 2 (1995) 157–170,
https://doi.org/10.1109/87.388124

[78] AT&T, Open Source Security Information and Event Management (SIEM) product,
https://cybersecurity.att.com/products/ossim

[79] SECURITY ONION SOLUTIONS, Security Onion 2 Latest version: 2.4.30, Security Onion, Evans, GA,
https://securityonionsolutions.com/software/

[80] KALEVA, H., "Electronic operating manual for NPP", paper presented at Int. Youth Nuclear Congress (IYNC2010), Cape Town, 2010.

[81] ELECTRIC POWER RESEARCH INSTITUTE, Implementation, Deployment, and Continuing Improvement of Electronic Work Packages, Rep. 3002018163, EPRI, Palo Alto, CA (2020).

[82] INTERNATIONAL ATOMIC ENERGY AGENCY, Safe Use of Smart Devices in Systems Important to Safety in Nuclear Power Plants, Safety Reports Series No. 111, IAEA, Vienna (2023).

[83] INTERNATIONAL ELECTROTECHNICAL COMMISSION, Nuclear Power Plants — Instrumentation and Control Important to Safety — Selection and Use of Industrial Digital Devices of Limited Functionality, IEC 62671:2013, IEC, Geneva (2013).

[84] SMITH, J.A., XU, C., DENG, Y., MANJUNATHA, K.A., AGARWAL, V., Wireless Sensing and Communication Capability from In-Core to a Monitoring Center, Idaho National Laboratory, Idaho Falls, ID (2020),
https://doi.org/10.2172/1668811

[85] KIM, T., AHN, M., KWON, J., LEE, J., "Application methodology of wireless communication technology for nuclear power plants", paper presented at Int. Symp. on Future I&C for Nuclear Power Plants (ISOFIC 2017), Gyeongju, Republic of Korea, 2017.

[86] ELECTRIC POWER RESEARCH INSTITUTE, Implementation Guideline for Wireless Networks and Wireless Equipment Condition Monitoring, Rep. 1019186, EPRI, Palo Alto, CA (2009).

[87] AGARWAL, V., BUTTLES, J.W., BEATY, L., NASER, J., Wireless online position monitoring of manual valve types for plant configuration management in nuclear power plants, IEEE Sensors J. **17** 2 (2017) 311–322,
https://doi.org/10.1109/JSEN.2016.2615131

[88] CHENG, H., et al., Evanescent-mode-resonator-based and antenna-integrated wireless passive pressure sensors for harsh-environment applications, Sens. Actuators A: Phys. **220** (2014) 22–33,
https://doi.org/10.1016/j.sna.2014.09.010

[89] INTERNATIONAL ELECTROTECHNICAL COMMISSION, Nuclear Power Plants — Instrumentation and Control Important to Safety — Use and Selection of Wireless Devices to be Integrated in Systems Important to Safety, IEC TR 62918:2014, IEC, Geneva (2014).

[90] NUCLEAR REGULATORY COMMISSION, Assessment of Wireless Technologies and Their Application at Nuclear Facilities, NUREG/CR-6882, NRC, Washington, DC (2006).

[91] MANJUNATHA, K.A., AGARWAL, V., Multi-band heterogeneous wireless network architecture for industrial automation, Wireless Personal Commun. **123** 4 (2022) 3555–3573,
https://doi.org/10.21203/rs.3.rs-163125/v1

[92] WON, S.M., et al., Multimodal sensing with a three-dimensional piezoresistive structure, ACS Nano **13** 10 (2019) 10972–10979,
https://doi.org/10.1021/acsnano.9b02030

[93] INTERNATIONAL ELECTROTECHNICAL COMMISSION, Functional Safety of Electrical/Electronic/ Programmable Electronic Safety-related Systems, IEC 61508, Parts 0 to 7, IEC, Geneva (2010).

[94] INTERNATIONAL ELECTROTECHNICAL COMMISSION, Functional Safety — Safety Instrumented Systems for the Process Industry Sector, IEC 61511, Parts 0 to 4, IEC, Geneva (2021).

[95] INTERNATIONAL ELECTROTECHNICAL COMMISSION, OPC Unified Architecture, IEC 62541, Parts 1 to 14, IEC, Geneva (2020).

[96] INTERNATIONAL ELECTROTECHNICAL COMMISSION, Programmable Controllers — Part 3: Programming Languages, IEC 61131-3:2013, IEC, Geneva (2013).

[97]   PLCOPEN, PLCopen web site, Gorinchem, Kingdom of the Netherlands (2024),
      https://plcopen.org/

[98]   OPC FOUNDATION, The OPC Foundation releases the OPC UA Field eXchange (UAFX) specifications, OPC
      Foundation News, Scottsdale, AZ (8 Nov. 2022),
      https://opcfoundation.org/news/press-releases/the-opc-foundation-releases-the-opc-ua-field-exchange-uafx-
      specifications/

[99]   OPC FOUNDATION, Digital transformation at Groupe Renault with GoogleCloud and OPC UA, OPC Foundation
      News, Scottsdale, AZ (22 Feb. 2022),
      https://opcfoundation.org/news/opc-foundation-news/digital-transformation-at-groupe-renault-with-googlecloud-
      and-opc-ua/

[100] INTERNATIONAL ORGANIZATION FOR STANDARDIZATION, INTERNATIONAL ELECTROTECHNICAL
      COMMISSION, Information Technology for Learning, Education and Training — Human Factor Guidelines for
      Virtual Reality Content — Part 1: Considerations When Using VR Content, ISO/IEC TR 23842-1:2020,
      ISO/IEC, Geneva (2020).

[101] INTERNATIONAL ORGANIZATION FOR STANDARDIZATION, INTERNATIONAL ELECTROTECHNICAL
      COMMISSION, Information Technology for Learning, Education and Training — Human Factor Guidelines for
      Virtual Reality Content — Part 2: Considerations When Making VR Content, ISO/IEC TR 23842-2:2020,
      ISO/IEC, Geneva (2020).

[102] INTERNATIONAL ORGANIZATION FOR STANDARDIZATION, INTERNATIONAL ELECTROTECHNICAL
      COMMISSION, Information Technology — Computer Graphics, Image Processing and Environmental
      Representation — Sensor Representation in Mixed and Augmented Reality, ISO/IEC 18038:2020,
      ISO/IEC, Geneva (2020).

[103] INTERNATIONAL ELECTROTECHNICAL COMMISSION, Energy Management System Application Program
      Interface (EMS-API), IEC 61970, All Parts, IEC, Geneva (2022).

[104] ELECTRIC POWER RESEARCH INSTITUTE, Application Integration Using Standards-based Messaging:
      Implementation of Energy Supply Common Information Model (ES-CIM) Interfaces, Rep. 3002020923, EPRI, Palo
      Alto, CA (2021).

[105] EUROPEAN NETWORK OF TRANSMISSION SYSTEM OPERATORS FOR ELECTRICITY, Common
      Information Model (CIM) for Grid Models Exchange, ENTSO-E, Brussels,
      https://www.entsoe.eu/data/cim/cim-for-grid-models-exchange/

[106] ENERGY CENTRAL, Interoperability Event Demonstrates Successful Grid Model Data Exchange (2022),
      https://energycentral.com/o/EPRI/interoperability-event-demonstrates-successful-grid-model-data-exchange

[107] INTERNATIONAL ORGANIZATION FOR STANDARDIZATION, Mechanical Vibration — Measurement and
      Evaluation of Machine Vibration — Part 8: Reciprocating Compressor Systems, ISO 20816-8:2018,
      ISO, Geneva (2018).

[108] INTERNATIONAL ORGANIZATION FOR STANDARDIZATION, Mechanical Vibration — Measurement and
      Evaluation of Machine Vibration — Part 3: Industrial Machinery with a Power Rating above 15 kW and Operating
      Speeds between 120 r/min and 30 000 r/min, ISO 20816-3:2022, ISO, Geneva (2022).

[109] MEEHAN, M., Where data goes to die: Big data still holds answers, but they're not where you're looking for them,
      Forbes (8 Dec. 2016),
      https://www.forbes.com/sites/marymeehan/2016/12/08/where-data-goes-to-die-big-data-still-holds-answers-but-theyre-not-
      where-youre-looking-for-them/

[110] MATHWORKS, MathWorks web site (2025),
      https://www.mathworks.com/

[111] INTERNATIONAL ORGANIZATION FOR STANDARDIZATION, Condition Monitoring and Diagnostics of
      Machines — Data Interpretation and Diagnostics Techniques — Part 1: General Guidelines, ISO 13379-1:2012,
      ISO, Geneva (2012).

[112] INTERNATIONAL ORGANIZATION FOR STANDARDIZATION, INTERNATIONAL ELECTROTECHNICAL
      COMMISSION, Information Technology — Artificial Intelligence — Artificial Intelligence Concepts and
      Terminology, ISO/IEC 22989:2022, ISO/IEC, Geneva (2022).

[113] PARK, J., KIM, T., SEONG, S., KOO, S., Control automation in the heat-up mode of a nuclear power plant using
      reinforcement learning, Prog. Nucl. Energy **145** (2022) 104107,
      https://doi.org/10.1016/j.pnucene.2021.104107

[114] WILES, J., What's New in Artificial Intelligence from the 2022 Gartner Hype Cycle, Hype Cycle for Artificial Intelligence (AI), Gartner, Stamford, CT (2022),
https://www.gartner.com/en/articles/what-s-new-in-artificial-intelligence-from-the-2022-gartner-hype-cycle

[115] INTERNATIONAL ELECTROTECHNICAL COMMISSION, Nuclear Facilities — Instrumentation and Control, and Electrical Power Systems — Artificial Intelligence Applications, IEC TR 63468:2023, IEC, Geneva (2023).

[116] INTERNATIONAL ATOMIC ENERGY AGENCY, Current Practices and Future Trends in Expert System Developments for Use in the Nuclear Industry, IAEA-TECDOC-769, IAEA, Vienna (1994).

[117] INTERNATIONAL ORGANIZATION FOR STANDARDIZATION, INTERNATIONAL ELECTROTECHNICAL COMMISSION, Software and Systems Engineering — Software Testing — Part 11: Guidelines on the Testing of AI-based Systems, ISO/IEC TR 29119-11:2020, ISO/IEC, Geneva (2020).

[118] INTERNATIONAL ATOMIC ENERGY AGENCY, Considerations for Deploying Artificial Intelligence Applications in the Nuclear Power Industry, IAEA Nuclear Energy Series No. NR-T-1.26, IAEA, Vienna (2025),
https://doi.org/10.61092/iaea.s6uy-wjt8

[119] INTERNATIONAL ATOMIC ENERGY AGENCY, Systematic Approach to Training for Nuclear Facility Personnel: Processes, Methodology and Practices, IAEA Nuclear Energy Series No. NG-T-2.8, IAEA, Vienna (2021).

[120] INTERNATIONAL ATOMIC ENERGY AGENCY, Knowledge Management and Its Implementation in Nuclear Organizations, IAEA Nuclear Energy Series No. NG-T-6.10, IAEA, Vienna (2016).

[121] INTERNATIONAL ATOMIC ENERGY AGENCY, Knowledge Loss Risk Management in Nuclear Organizations, IAEA Nuclear Energy Series No. NG-T-6.11, IAEA, Vienna (2017).

[122] INTERNATIONAL ATOMIC ENERGY AGENCY, Computer Security Approaches to Reduce Cyber Risks in the Nuclear Supply Chain, IAEA, Vienna (2022).

MONITORING AND DIAGNOSTICS OF INDUSTRIAL THERMODYNAMIC PROCESSES WITH METROSCOPE REAL TIME APPLICATION

I–1. INTRODUCTION

Metroscope is a company founded in 2018 that develops a commercially available software application for real time diagnostics of industrial thermodynamic processes, including nuclear power plants. These diagnostics are performed by leveraging real time operational data, past operating experience and artificial intelligence (AI). Though all current uses concern thermal performance, other performance factors (e.g. environmental performance, flexible operation or cogeneration performance) could be addressed, provided the phenomena and deteriorations that can affect them can be modelled.

The Metroscope application relies on:

— A thermodynamic model that simulates both the nominal and faulty behaviours of the plant;
— An inferential engine that identifies process faults (leaks, fouling, etc.) from gaps between the model and observed reality.

The Metroscope application has so far been deployed on the steam circuit of 68 power plant units around the world. Among those are all the 56 active units of the French nuclear fleet. Approximately 90% of faults identified by Metroscope have been confirmed by on-site engineering teams.

This annex describes the challenges addressed by the Metroscope application, presents its AI principles and provides examples of fault identification obtained when processing live nuclear power plant data.

I–2. CHALLENGES ADDRESSED

Monitoring and diagnostics in industrial thermodynamic processes (i.e. the identification of faults impacting plant thermal performance and/or component reliability in real time) requires abundant data, advanced expertise, accurate modelling and significant effort. The time needed for these root cause analyses can be significant and results in prolonged operation at reduced plant performance while awaiting fault identification and subsequent corrective actions. Also, any false positive diagnosis can exacerbate the time plants operate at reduced thermal performance, resulting in additional lost revenues.

It is common for plants to operate with multiple, simultaneous performance-impacting (but not safety-impacting) issues (e.g. leaking or misaligned valves, circuit isolation, internal leaks in heat exchangers, heat exchanger fouling, abnormal tank water level). To diagnose those issues, they use hundreds of instruments installed at locations specified by the original design or by subsequent modifications.

The abundance of data generated by so many sensors and time stamps exacerbates the complexity of 'manual' cause analysis. Additional complexity is due to measurement uncertainty and drift, sometimes due to changes in plant load, process state and environmental conditions.

The main challenge addressed by Metroscope is the following: assisting plant engineers in identifying faults using software that elaborates sensor gaps automatically through digital twinning, delivering real time cause diagnostics and user friendly outputs (consistent with the concept of 'explainable AI'). This challenge is successfully achieved by the application. Additional challenges tackled are discussed in the paragraphs below.

Plant personnel may not be able to pinpoint a fault occurring at a plant component by looking just at measurements from nearby sensors without support from a full system model, due to those sensors being limited in number. This may lead to the installation of additional sensors (e.g. ultrasonic flowmeters), which results in greater maintenance costs, including expenses for installation and routine calibration. In contrast, the Metroscope application extracts augmented information from sensors already available by running in real time a thermohydraulic digital twin model of the full system, thus obtaining process parameters as if 'virtual' sensors were installed. This enables the observation of local process conditions more thoroughly than simpler data visualization methods.

An additional challenge for plant engineers occurs when faults are identified. When presenting their findings to management or maintenance teams, engineers are often asked for a justification of their analysis and an estimate of the fault's impact on power production or plant operability. These requests lead to additional burden on the schedule that the Metroscope application eases by empowering the engineers with automatic diagnosis and impact assessment. This impact quantification is especially important when multiple performance improvement actions are carried out simultaneously; it is then often difficult to demonstrate the effectiveness and impact of individual actions. The Metroscope application provides this information automatically.

Finally, the capability to detect small faults before they degrade into large faults is a key challenge. Most faults tend to start at a level detectable with a digital twin model before reaching a level detectable by traditional measurements and calculations. By this time, they may already have caused important performance losses and/or component damage requiring costly maintenance. The lower detectability threshold of a finely tuned, full system digital twin like those used by Metroscope compared to a 'manual' or single component analysis is due to reliance on information and statistics from many more sensors, as has been demonstrated in several real life cases.

I–3. METHOD OVERVIEW

The three key terms used here to describe the Metroscope method are 'symptom', 'diagnosis' and 'fault'. A medical analogy can help define them: by analysing some measurements (e.g. a blood test), a doctor identifies symptoms (e.g. low red cell count) and then makes a diagnosis to explain those symptoms (e.g. vitamin B12 deficiency). Here, the diagnosed cause of a set of symptoms is referred to as a fault.

The Metroscope method follows similar steps:

(1) Sensor data are read.
(2) Plant symptoms are identified based on sensor readings, taking account of non-symptomatic signal drifts (e.g. due to changes in process state, plant load or environmental conditions around the component).
(3) The cause of symptoms is identified as a set of specific faults, taking account of systematic and random uncertainty.
(4) The impact of the identified faults on system performance (e.g. plant power output) is quantified.

To perform steps 2 and 3 above (i.e. to identify symptoms and faults), the Metroscope method relies on two software engines, both represented in Fig. I–1: the digital twin, described in Section I–4, for a plant steam circuit application, and the inferential engine, described in Section I–5. As Fig. I–1 shows, the digital twin is used in the Metroscope method for two key purposes:

— To compute the normal, expected behaviour and thereby detect symptoms by comparing it with the observed behaviour;
— To compute behaviours under various fault conditions.

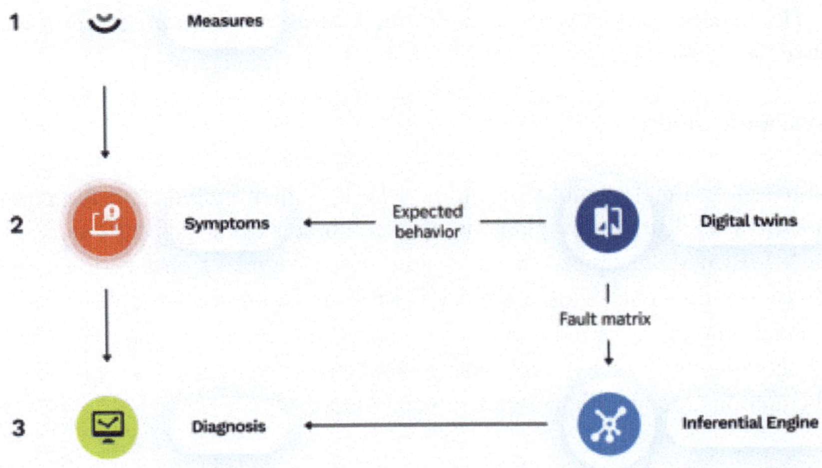

FIG. I–1. Metroscope diagnosis method (courtesy of Metroscope, France).

The inferential engine performs the diagnosis using a Bayesian network to compare the observed symptoms with those of a library of faults generated by the digital twin. This step considers very large numbers of combinations of possible faults with different magnitudes. The magnitude quantification depends on the fault's nature. For instance, a leak is quantified by its mass flow rate. The combination of active faults and associated magnitudes with the highest confidence level is selected and presented to users.

I–4. DIGITAL TWIN

While the Metroscope method described in Section I–3 is applicable to different types of processes, this section describes the digital twin used in the most common Metroscope applications: the steam circuit of a power plant.

An ideal thermodynamic process functions without faults. The Metroscope digital twin outputs in this operation mode are the expected behaviour shown in Fig. I–1 and determine the values that should be measured by sensors (mostly temperature, pressure and mass flow rate), given a set of operating conditions and external conditions (hereby called boundary conditions). When faults affect the process, sensor readings can differ substantially from the expected behaviour; the differences exceeding an uncertainty threshold constitute a set of symptoms.

Each digital twin is made of discrete blocks corresponding to plant components. Each contains zero dimensional sets of equations (typically mass and energy conservation equations), and the network of connected blocks describes the system. Metroscope implements the blocks in a library written in the Modelica language: the open source Metroscope modelling library (MML).

Modelica is an acausal equation language. Acausal means that, for example, a model can be used in a 'direct mode' to compute the values of sensors knowing the boundary conditions and parameters of all system components. Nonetheless, the model can be seamlessly reversed to compute the component parameters taking sensor values as inputs (this mode is called 'calibration'). The Modelica language is highly suited for system modelling, since it is object oriented and allows for the definition of packages built from other packages without limits in the number of nested items. Most Modelica development environments offer drag and drop graphical user interfaces to quickly build full system models from component libraries.

The currently used Metroscope digital twin models are based on steady state equations. Unlike other types of models (dynamic simulators, 3-D turbine models, etc.), the models used by Metroscope can easily handle computation of a stable thermodynamic process with a high level of complexity (e.g. over 100 components). The benefits of this type of model for Metroscope's purposes are low computational cost and reasonable development cost.

I–4.1. Thermohydraulic model

The MML is based on the ThermoSysPro library [I–1], which includes all the components found in the steam circuit of a nuclear power plant or of a combined cycle gas turbine. Figure I–2 shows an example of the MML modular approach for a simplified secondary circuit. Each line represents a connection between two components, and each block is a process component (valves, turbines, generators, heat exchangers, condensers, etc.).

In all these components, mass and energy balance equations are specified, as well as phenomenological equations for specific components, including:

— Stodola's law in turbines, which links flow rate to pressure in the turbine using a specific coefficient;
— Number of transfer units–effectiveness (NTU–epsilon) heat exchange laws, which compute the power exchanged in a heater based on the inlet temperatures, geometry and heat transfer coefficient;
— Pump characteristics.

These equations are embedded in the library. The model developer then connects the components together and specifies the inputs and outputs of the model, bearing in mind that the total system of equation has to be square for the compiler to solve it.

Models are then exported and deployed in the Metroscope application using the Functional Mock-up Interface (FMI) standard [I–2].

I–4.2. Nominal model

The nominal model computes expected sensor values based on a few boundary conditions. Because the phenomenological laws contain complex parameters (heat transfer coefficients, friction loss coefficients, efficiencies, etc.), the model has to be calibrated so that the results are accurate. Model calibration can be done following two principles:

— Design based: The parameters are computed from the design characteristics of the component.
— Data based: The parameters are fitted so that the results of the model match a given data set from the actual process.

For performance evaluation, both approaches have advantages and disadvantages, so the Metroscope method combines them in an optimal trade-off. The design based approach gives a baseline that is the best performance imaginable for the plant but is not realistic for older assets, which likely have drifted from design through decades of operation. This baseline is, therefore, not currently achievable and is of little use to plant operators. The data based approach, on the other hand, gives a baseline that matches current behaviour, which may unintentionally mask possible faults in the process. The resulting baseline for such a calibration may not be optimistic enough to be representative of the best possible performance. Therefore, the Metroscope calibration combines the two approaches to extract parameters corresponding to realistic best plant behaviour.

As mentioned above, calibration is performed using the acausal nature of Modelica models. Consider the simple example of an electrical resistor. Voltage (V) and current (I) are linked by Ohm's law, $V = R \times I$, where R is resistance. A direct approach assumes that R is known and V is a measured boundary condition. The model can then compute I, which can then be compared to the measured I to identify any symptoms. In a reverse approach, V is still established by measurements, but the goal is to obtain an empirical value of R. Therefore, a measured value for I representative of operation at high resistor

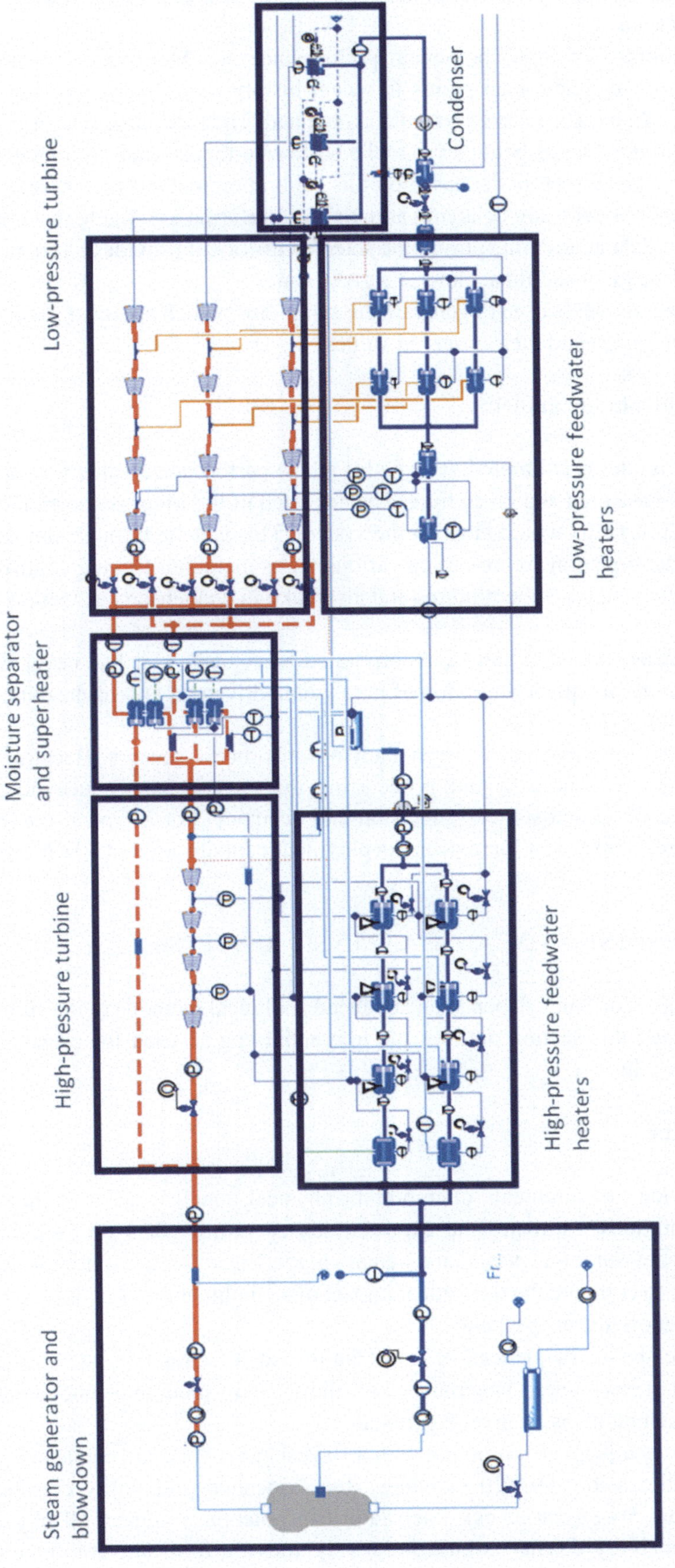

FIG. I–2. *Example of a Modelica model based on the Metroscope modelling library (courtesy of Metroscope, France).*

performance is entered as an input. The model then gives a value of R to be used in the direct digital twin model for diagnostics purposes.

In practice, to calibrate for high component performance, the Metroscope method evaluates the reverse model with all the available data points in recent history (e.g. 3–5 years). For each parameter needing calibration (e.g. R), a distribution over time is obtained. The modelling engineer can then extract data from these distribution values as being close to the best performance each component has effectively reached in recent history. The direct model based on this calibration method will therefore operate under the assumption that all components are behaving at their best performance. While this condition may not be reflected in the actual data at any time during the selected history, it provides a reliable representation of the performance that the plant should achieve at a given time.

The calibrated direct model is used to generate the symptoms, which are the differences between the nominal expected sensor values and their observed values.

I–4.3. Introducing faults in the model

Once the nominal model is calibrated, it provides a best performance value for each sensor, given the boundary conditions measured at a given time. The next step of the Metroscope method computes the impact that a specific set of faults would have on the system. This is done through simulation by writing a fault model that includes equations representing various faults including leaking control valves, bypass valves, fouling of heat exchangers, tube ruptures and air intake in condensers, as well as feedwater flow rate bias (see Fig. I–3).

Any fault that can be modelled can be incorporated here, even if it has never happened at the power plant. This overcomes a typical limitation of purely data driven AI approaches like those based on advanced pattern recognition.

A fault signature is the impact of a given fault on all process sensors. Because it looks at the scattered impacts even far from where the fault is occurring, this method can produce more complete fault signatures, resulting in good diagnosis with precision and accuracy. For a typical nuclear power plant secondary circuit, the fault library of a Metroscope deployment contains around 80 different faults.

I–5. INFERENTIAL ENGINE AND CAUSE ANALYSIS RESULTS

The accurate diagnosis of faults depends on advanced analytical methods to interpret plant data and present actionable insights; this section outlines the inferential engine used for cause analysis and the approach to displaying results.

I–5.1. Inferential engine

Plant sensors provide measurements of important physical quantities. The digital twin, fed with measured boundary conditions, estimates the expected sensor values when no faults are active. The symptoms are the gaps between the expected and actual values. Since the faults are also modelled in the digital twin, it is possible to compute the theoretical impact of a combination of faults and compare it with current symptoms to diagnose the active faults.

The symptoms are the consequences of both faults and uncertainty (the latter resulting from a combination of sensor measurement uncertainty and digital twin modelling uncertainty). Therefore, finding the cause of the symptoms is an inverse problem.

Metroscope addresses this inverse problem with a Bayesian network and can hence estimate, given the symptoms (typical dimension: 80–120 variables), the associated probability density of the faults' values (typical dimension: 50–80 variables). Each fault can potentially impact all physical quantities. The estimate of this probability density is built iteratively, moving from one combination of faults to

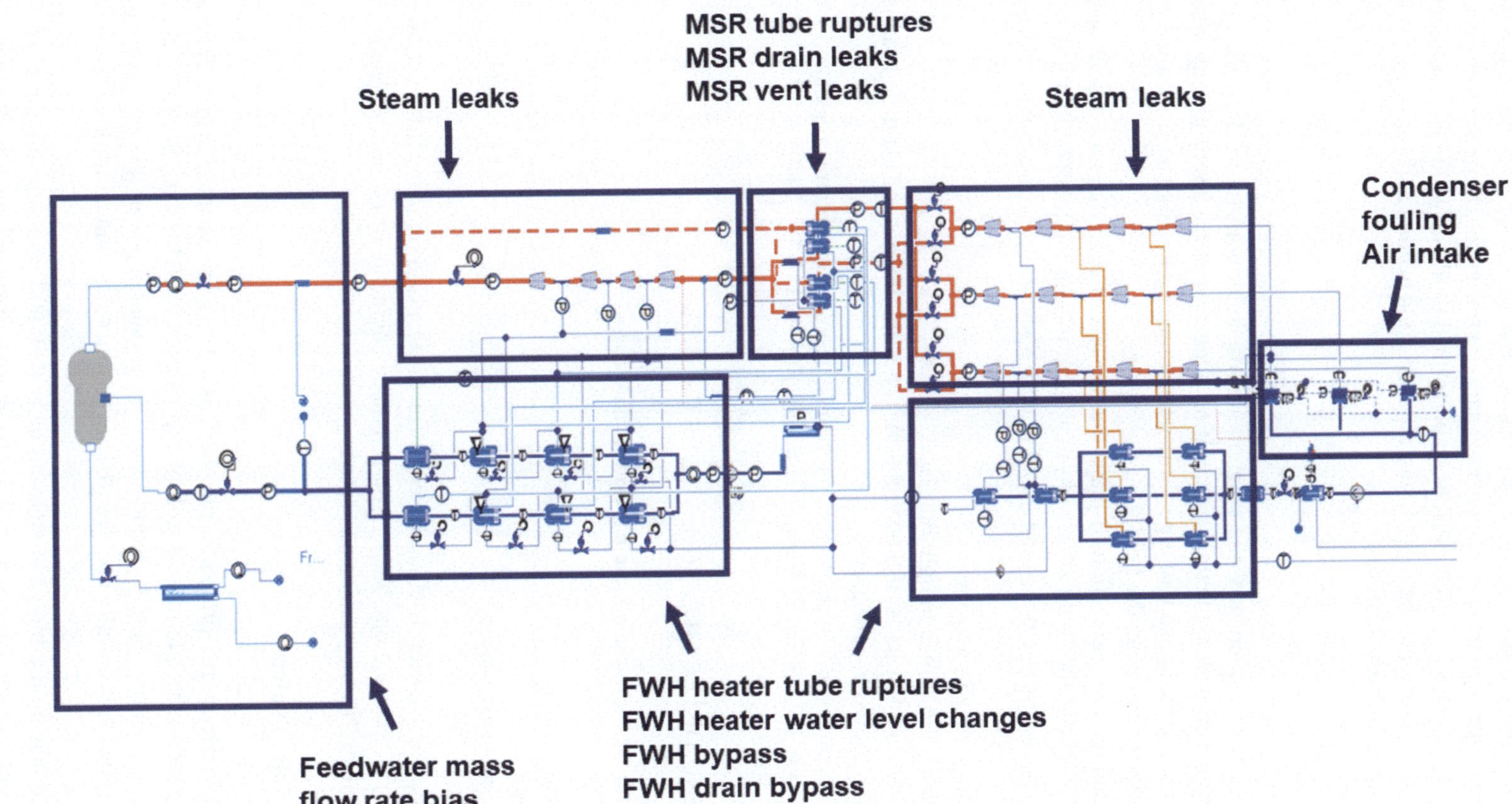

FIG. I–3. *Modelica digital twin based on the Metroscope modelling library, showing some of the faults included in the model. FWH — feedwater heater; MSR — moisture separator reheater. (Courtesy of Metroscope, France.)*

another. The higher the likelihood of a combination of faults, the higher the number of iterations in the associated area.

At each iteration, a combination of faults is propagated in the symptoms space. The computation of the impact of the combination of faults relies on the digital twin described in the previous section. The computation time for a single simulation of such a digital twin varies from a few seconds to a few minutes. Since thousands of such computations need to be performed to identify the best combination of faults, surrogate models that approximate the digital twin outputs are generated and called instead of the full digital twin, to enable quicker cause analyses.

I–5.2. Displaying results

Based on real time plant measurements, the combination of faults with the highest likelihood is provided to the user and displayed with contextual information. One goal of the application is to provide a detailed explanation of the diagnostic, so that a performance engineer can validate or invalidate it, for instance, with additional information from the field.

To facilitate understanding, the application displays the impacts of each diagnosed fault on power loss (left histogram in Fig. I–4) together with all the physical quantities in the model (right histogram in Fig. I–4, that is normalized by acceptable uncertainty, σ). Figure I–4 shows a real life, anonymized case:

— Fault 'Tube_Rupt_1' is shown in magenta.
— The associated symptoms are shown in light blue.
— The residuals (i.e. the remaining parts of the symptoms, once the impacts of identified faults have been accounted for) are shown in dark grey.
— It is evident that the diagnosis of 'Tube_Rupt_1' is mainly based on symptoms associated with the physical quantities 'Flow_17', 'Flow_15', 'Flow_14' and 'Flow_13'.

I–6. SOME INTERESTING DETECTIONS (REAL USE CASES)

This section presents three anonymized nuclear power plant cases:

— A leaking circuit isolation valve;
— A reheater tube rupture;
— Condenser underperformance tracking.

Because the Metroscope application is cloud based, it can be accessed by many different users at a given utility (ranging from maintenance operators to performance engineers or corporate management). Therefore, the decision making efficiency is also presented.

I–6.1. Leaking circuit isolation valve

A retrospective health study was performed for the plant, based on a few years of historical data. The Metroscope application pointed out an intermittent leak of approximately 20 kg/s at a normally closed air operated valve, causing a high energy leak to the condenser. The associated power loss was estimated by the application at 2 MW(e). The plant initially decided to delay maintenance and monitor the evolution of the issue with the application.

Three months later, the detection rapidly increased to more than 60 kg/s, corresponding to a power loss of 7 MW(e). At that point, the site triggered field investigations that confirmed the leakage of high energy water through the valve identified by Metroscope (see Fig. I–5), while the valve always appeared closed from the control room. The site decided to close the valve manually.

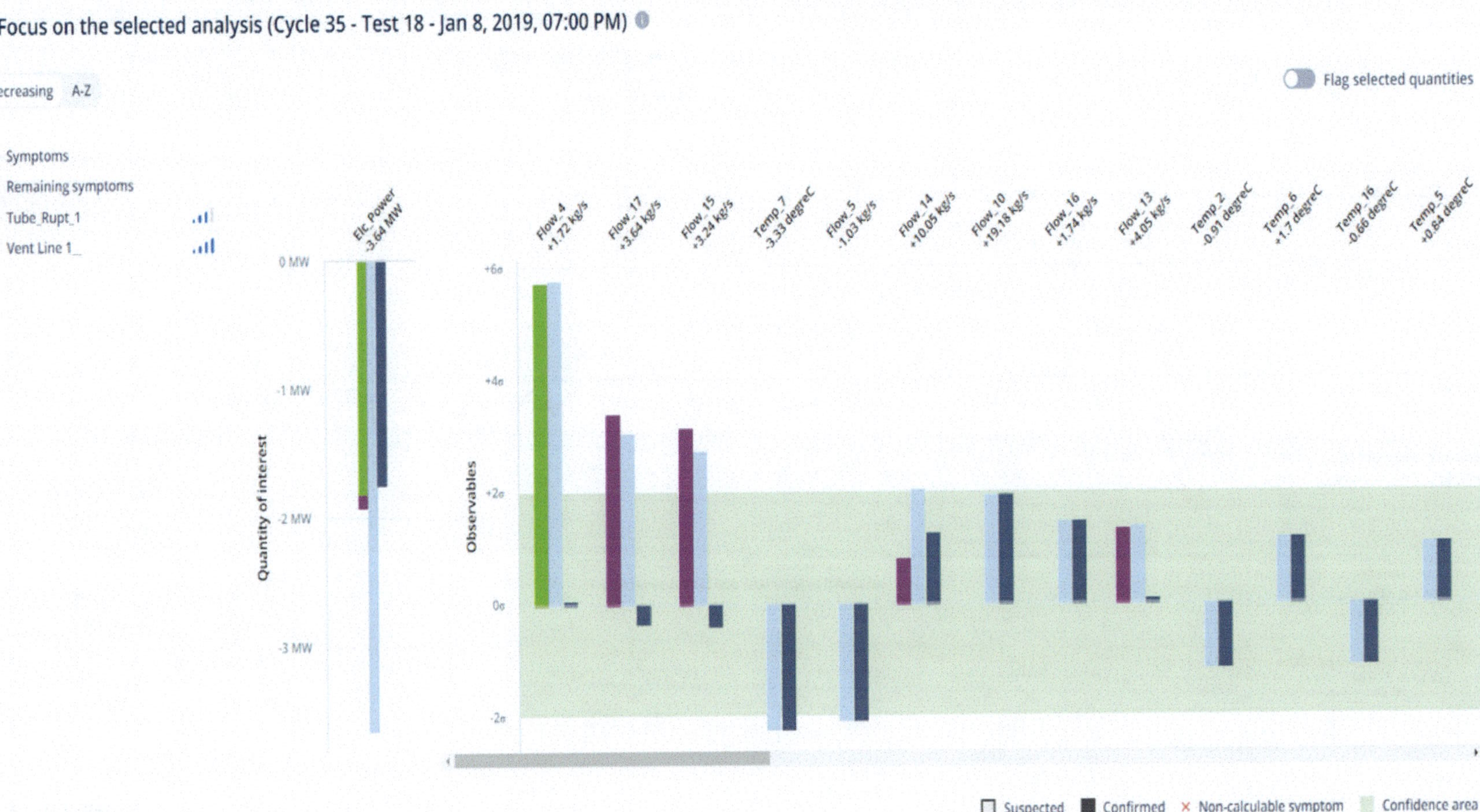

FIG. I–4. Display of a combination of faults and the associated impacts, at a given time stamp (courtesy of Metroscope, France).

FIG. I–5. Screenshot of the Metroscope summary page showing the use case. The valve leakage described in this case corresponds to the pink line. (Courtesy of Metroscope, France.)

After the action, the application still detected a 2.5 MW(e) loss associated with this valve, indicating that the maintenance operation had been only partially effective. Based on this evaluation, a second intervention was conducted to close the valve. After this second operation, the leak disappeared from the application results, indicating that the valve had been fully closed. Three months later, a planned outage allowed the valve to be opened and the diaphragm to be fixed, which was the cause of the valve malfunction.

Since there was no sensor providing direct information about the valve (neither mass flow rate nor downstream temperature), the fault was difficult to detect manually. The Metroscope application detected the leak by sensing multiple low impact effects generated by the fault on other physical quantities far away from the valve. The application also reported on the amplitude of the leak, which allowed the site to monitor it and choose the right time for maintenance.

At the time of diagnosis, the plant was experiencing at least three other faults, which were also identified and quantified by the application.

As the valve would have been leaking for a total of 19 months between the first detection and the plant outage, with a mean loss of 2 MW(e), the Metroscope application estimated that the total production loss would have been around 30 000 MWh without its diagnostics.

I–6.2. High pressure reheater tube rupture

Tube ruptures, especially when occurring on high pressure heat exchangers, are tracked for regulatory reasons but also to avoid component degradation. There are thousands of tubes per heat exchanger and they are not accessible during continuous operation. To detect a tube rupture, it is therefore necessary to monitor the impacted process parameters that are recorded by sensors. Detection through monitoring of surrounding parameters is usually only possible for significant ruptures involving multiple tubes. In this case, the Metroscope application was able to reduce the detection threshold by a factor of three.

According to operator statements, the typical detection threshold at this site was around 9 kg/s. Using Metroscope, the tube rupture was detected at a magnitude of 3 kg/s and later confirmed by the site. This detection took place approximately 10 weeks before the plant was able to see it without Metroscope diagnostics (see Fig. I–6).

I–6.3. Condenser underperformance tracking for better maintenance decisions

Condensers are key components for power plant performance and are therefore good examples of components closely monitored to prevent plant efficiency losses.

The top panel of Fig. I–7 shows the difference between the expected vacuum pressure in a condenser (green line) and the measured pressure (blue line). In the bottom panel of the figure, the difference between the two (the symptom) is plotted and normalized; the $\pm 2\sigma$ uncertainty band, shown in green, is ± 1 mbar. A steadily increasing symptom appears, with a significant increase around November 2019.

Figure I–8 shows the condenser losses (in percentage of fouling) and the megawatt losses (blue line) computed by the application. The latter was a constant 5 MW(e), then reached 19 MW(e). Following this increase, the thermal performance team decided to launch a field investigation and identified a malfunctioning vacuum pump. The decrease of the losses on the last points of Fig. I–8 corresponds to different maintenance actions that were quantified by the Metroscope application.

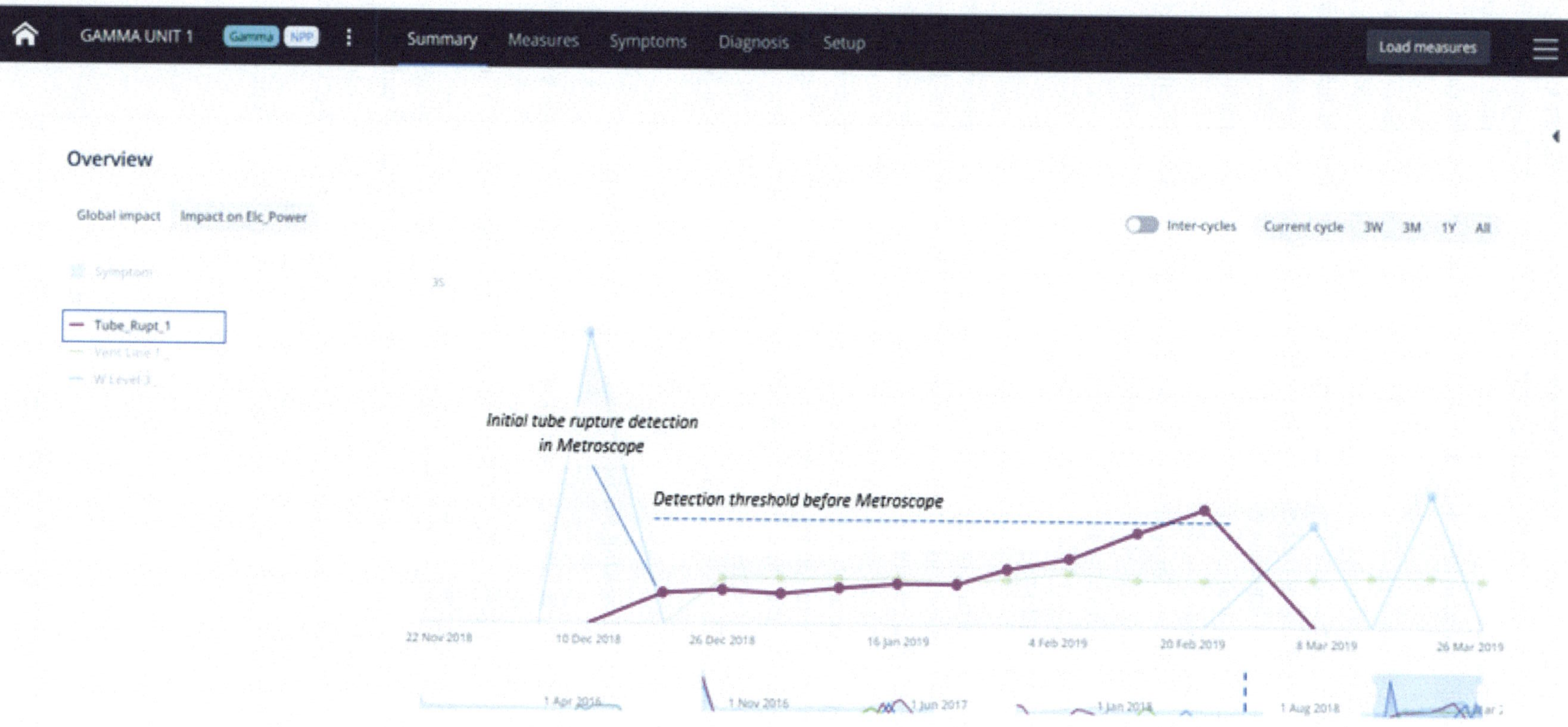

FIG. I–6. *Metroscope summary page showing the tube rupture use case (purple line) (courtesy of Metroscope, France).*

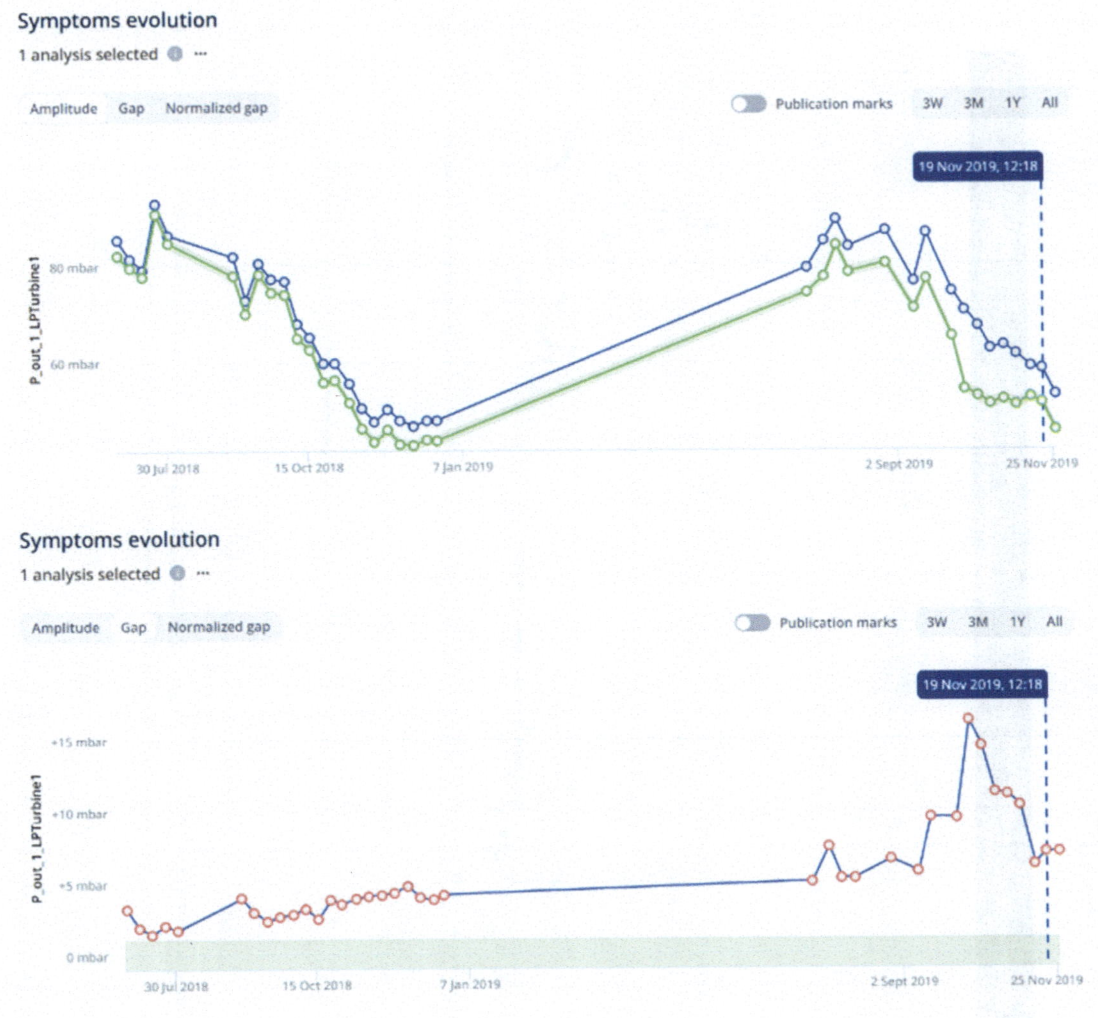

FIG. I–7. Evolution of the steam turbine back pressure (mbar). Top: Evolution of expected and measured values. Bottom: Evolution of the symptom. (Courtesy of Metroscope, France.)

I–7. CONCLUSION

This annex describes a digital twin method for performing diagnostics on complex power plants that is currently deployed on 68 nuclear power plant units worldwide. Its high reliability is demonstrated by manual confirmation by plant engineers of nearly 90% of diagnosed faults.

The technology achieves this reliability by combining a thermodynamic digital twin of the secondary circuit, including nominal and faulted behaviour, with a diagnosis algorithm based on a Bayesian network. This combination improves the treatment of large numbers of sensors for root cause analysis. It also enables fault impact quantification and maintenance action efficiency.

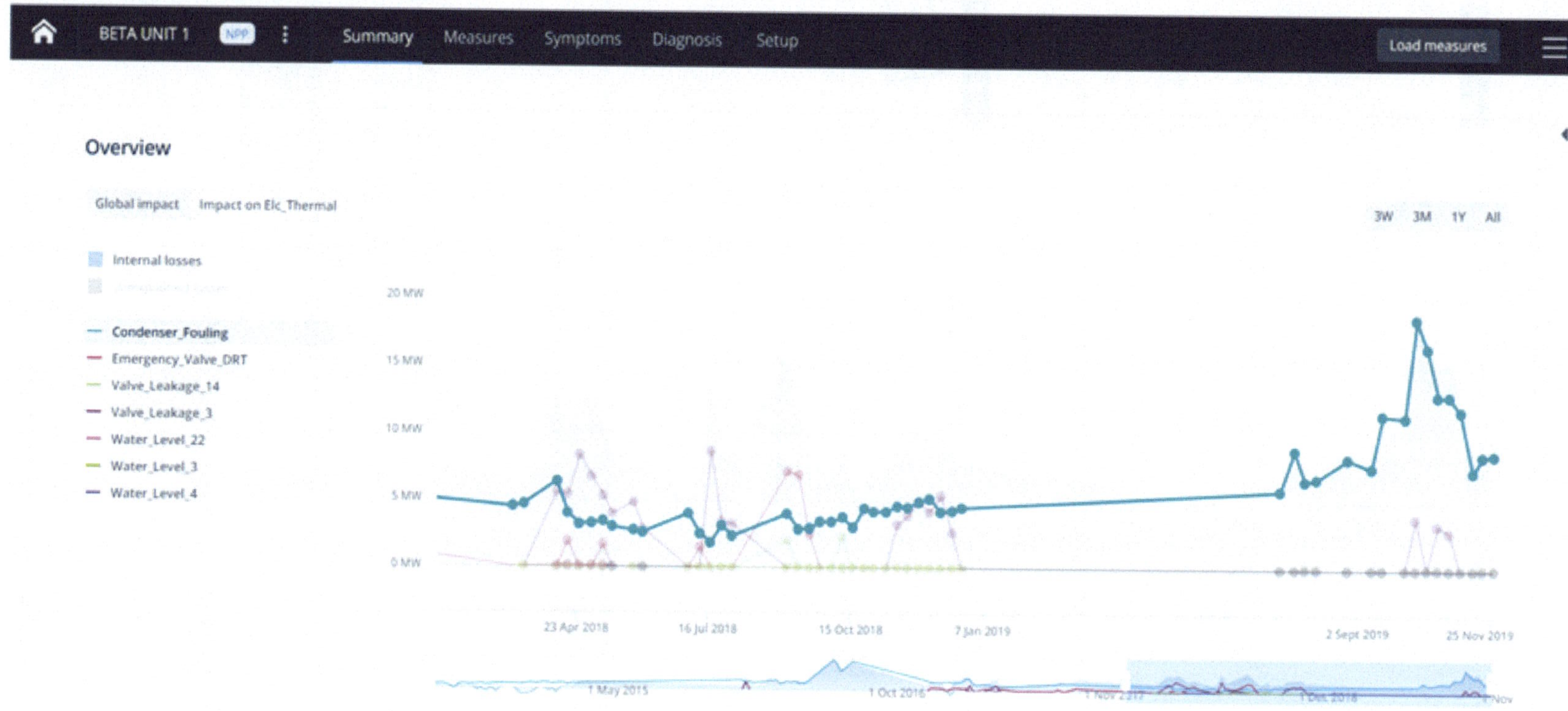

FIG. I–8. *Evolution of the impact of the condenser underperformance on the electric power output (courtesy of Metroscope, France).*

REFERENCES TO ANNEX I

[I–1]   EL HEFNI, B., BOUSKELA, D., Modeling and Simulation of Thermal Power Plants with ThermoSysPro: A Theoretical Introduction and a Practical Guide, Springer Nature Switzerland AG, Cham (2019), https://doi.org/10.1007/978-3-030-05105-1

[I–2]   MODELICA ASSOCIATION, Functional Mock-up Interface (FMI) Specification, Version 3.0, Modelica, Munich (May 2022), https://fmi-standard.org/docs/3.0/

Annex II

DATA DRIVEN DECISIONS WITH CADIS

II–1. INTRODUCTION

Framatome's lifetime management department offers products and services for condition monitoring, diagnostics and asset management. They include monitoring solutions for the reactor coolant system based on vibration, acoustic and leakage monitoring. These solutions are requested in some markets. They also include monitoring and diagnostics solutions for rotating machinery like pumps and non-rotating components, such as pipes/heat exchanges, for motor operated control valves and finally for fatigue of mechanical structures.

To improve the efficiency of its monitoring and diagnostics systems and reduce the engineering costs and effort, Framatome has developed a data analytics platform solution called CADIS [II–1], which is currently used in two ways: internally at Framatome as a stand alone system and externally in power plants in combination with existing data loggers and monitoring systems (see Fig. II–1).

CADIS is a configurable data driven decision support solution that automatically transforms collected equipment/engineering data into structured diagnostics information and analytics insights for decision making and plant asset optimization, using robotic process automation (RPA) and signal/text/language processing. After that, analytics insights like anomaly detection, fault pattern recognition and remaining useful life (RUL) estimation are derived using analytics models or artificial intelligence (AI) (see Ref. [II–2]). Finally, all results are put at the disposal of engineering teams in the form of customizable apps, dashboards or reports, helping them to make faster and more reliable decisions.

This annex describes a problem to be solved by plant operators, presents CADIS as a solution to the problem and shows some CADIS use cases with real plant asset data.

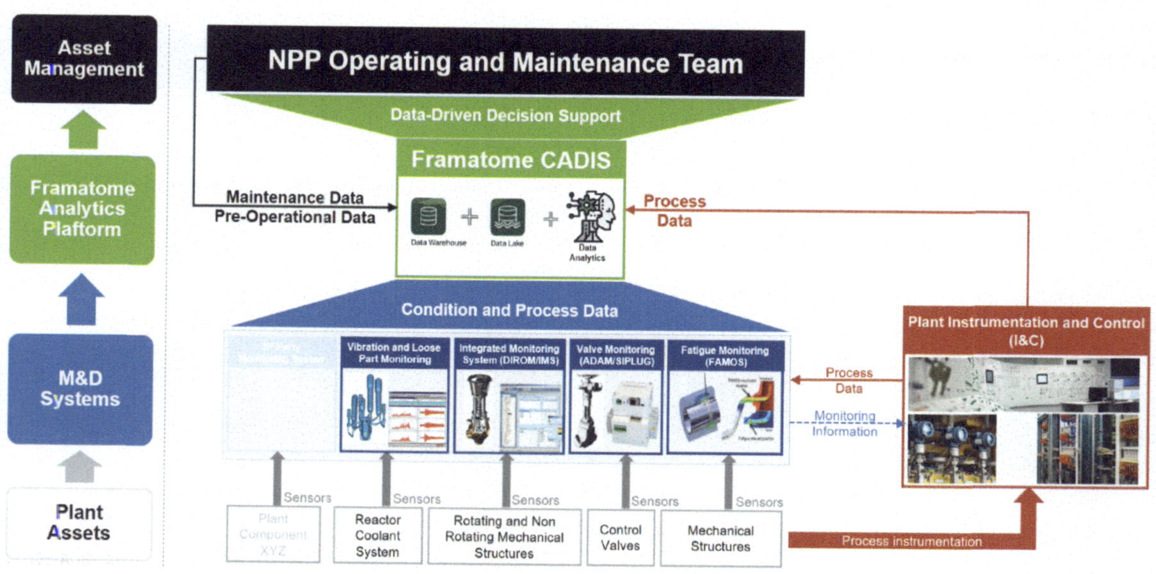

FIG. II–1. CADIS, a Framatome solution for data driven maintenance and asset management. M&D — monitoring and diagnostics; NPP — nuclear power plant. (Courtesy of F. Fomi Wamba, Framatome, Germany.)

II–2. PROBLEM TO SOLVE

In the past years, on-line monitoring in nuclear power plants was mostly done for critical equipment like turbines and main coolant pumps. Collected data at that time were reduced to fault indicators (e.g. effective and peak values) identified by standards (e.g. Refs [II–3] to [II–9]) or expert knowledge. Furthermore, the equipment data (condition, process and maintenance data) were stored separately and analysed by different experts or teams. Finally, condition monitoring was mainly focused on safety and not necessarily on plant availability and cost optimization.

Today, overall performance and reduction of operational costs are key objectives. With decreasing costs for instrumentation, condition monitoring hardware and computing systems, condition monitoring is no longer restricted to safety equipment and is extended to operational equipment to optimize maintenance intervals, maintenance duration, spare parts management and component replacement.

The increasing numbers of instrumented assets and the resulting huge amount of acquired plant data strengthen the demand for big data analytics solutions to efficiently store and convert data into analytics insights for smarter and faster decision making. In addition to that, the demand to move from curative and/or preventive maintenance to data driven maintenance like condition based maintenance and predictive maintenance using asset management tools is continuously rising, since data driven maintenance is a key measure for identifying plant issues before they occur or at least before there is a significant impact on plant reliability.

Framatome, as one of the leading suppliers in instrumentation and control as well as monitoring and diagnostics solutions in nuclear power plants, has contributed to this area for more than 40 years. CADIS, the development of which started several years ago, is the Framatome solution for the above mentioned problem.

II–3. CADIS BENEFITS TO PLANT PERFORMANCE

CADIS is a digital tool based on RPA and AI that helps engineering teams to achieve better work quality and results with less time and manual effort, by the collection, processing and transformation of huge amounts of engineering data from various sources — such as the core monitoring system (CMS), instrumentation and control, and computerized maintenance management system (CMMS) — and in various formats, including structured data (e.g. CSV or SQL databases, JSON, XML) and unstructured data (e.g. audio, images, videos, text).

The collection and processing of engineering input data is commonly done interactively in several sequential steps (process tasks) that require import and export or copy and paste by hand or by simple to complex processing scripts. Most of these engineering tasks are repetitive, time consuming and susceptible to human error. With the RPA methodology implemented in CADIS, simple repetitive tasks can be automated, allowing operators to focus on complex or high value adding tasks.

The coupling of RPA with AI facilitates the creation of advanced data driven decision support systems. While RPA focuses on data collection, preparation and processing, AI is responsible for analytics insights derivation (e.g. anomaly detection or forecasting, fault pattern recognition, RUL estimation using statistics, machine learning and deep learning for more reliable decision making).

Such information is useful for plant operators to understand, operate and maintain plant equipment better, and make faster, more reliable decisions regarding maintenance planning and asset management. The analytics insights derived from the collected data allow the plant operator to move from curative or scheduled maintenance to data driven maintenance. Data driven maintenance strategies in comparison to traditional time based maintenance can reduce the operational costs by optimizing the overall effectiveness (availability, reliability and maintainability) of instrumented plant assets.

II–4. DEPLOYMENT OF CADIS

For an efficient deployment of CADIS in a given nuclear power plant component, it is necessary to:

— Identify the critical faults of the component and corresponding measurements/features for detection and identification.
— Define and/or enhance the strategy for collecting these measurements/features.
— Configure an RPA+AI data analytics solution for the component.

II–4.1. Identification of critical component faults and their related monitoring and diagnostic features

According to ISO 13379-1 [II–1], plant operators who seek to implement continuous on-line monitoring for a specific component need to evaluate each significant degradation mechanism that affects that component. Then they can create a condition monitoring and diagnostic solution for that component as a combination of sensors, monitoring means and analytical methods to address the identified component degradation.

The idea behind condition based maintenance and predictive maintenance is 'don't fix anything that isn't broken', and take advantage of the following:

— There are warning signals before actual damage occurs.
— A failure mode produces symptoms detectable by suitable monitoring techniques, resulting in a high diagnosis and prognosis confidence.
— Increased confidence in diagnostics and prognostics can be gained by using correlation/fault pattern recognition techniques (data driven diagnostics) and/or knowledge based diagnostics (human experience).

The aim of failure mode and symptoms analysis (FMSA) is to select monitoring technologies and strategies that maximize the confidence level in the diagnosis and prognosis of any given failure mode. In order to find an optimal monitoring and diagnostics strategy for a nuclear power plant component, it is better to carry out the FMSA with nuclear power plant maintenance teams responsible for that component. Key aspects to address include:

— What are the functions and failure modes of the component?
— What are the causes and consequences of the critical failure modes?
— Which monitoring/diagnostic information can be used to detect/identify the critical failure modes?
— Build a component fault matrix as a relationship between component faults and diagnostic features (see Table II–1).

TABLE II–1. COMPONENT FAULT MATRIX ASSOCIATING DIAGNOSTIC FEATURES (CONDITION INDICATORS) AND COMPONENT FAULTS

Component fault	Diagnostic features		
	Feature 1	Feature 2	…
Fault 1			
Fault 2			
…			

II–4.2. Definition and optimization of the data collection strategy

CADIS, as a data analytics solution for predictive maintenance, relies on the following data.

— Sensor or measurement data:
  - Dynamic measurement or high resolution data, also called sound data, such as vibration signals;
  - Static measurement data or low resolution data, also called structured data, such as operating/process parameters (e.g. temperatures, flows, pressures).
— Component design data: Information such as natural frequencies, kinematics data, bearing types, motor types, list of failure modes.
— Component history data: Records of the fault history, operational history and maintenance history of the component.

The above mentioned data are often stored across diverse data platforms, including the CMS, programmable logic controller, supervisory control and data acquisition system, CMMS and equipment design documentation. For optimal data driven maintenance of a given plant component, it is necessary to:

— Collect all the required measurement data identified by the component fault matrix.
— Collect all process data related to the operation/operating conditions of the component.
— Collect all operational logs and/or maintenance data related to the operation of the component.
— Store all collected data in one central place (e.g. the CADIS server) and in an open format.

II–4.3. Configuration and deployment of CADIS for a nuclear power plant component

The aim of this step is to configure and deploy CADIS to do the following (see Fig. II–2):

— Collect all the available equipment data (condition, process, maintenance), also called 'raw data', at one central place (data collection) using industry standard communication interfaces.
— Transform 'raw data' into structured time series information or 'feature data' (data processing) using signal/text/natural language processing methods.
— Store 'feature data' in a time series database (historian).
— Convert 'feature data' into meaningful insights using data analytics models (data analytics).
— Put all data (raw data, feature data and analytics data) at the disposal of engineering teams in the form of customizable status reports or dashboards (data delivery).

II–4.3.1. CADIS data collection

The task of the data collection module is to collect all the available equipment data (condition and process, maintenance and design data) from connected data sources using standard communication interfaces at one central place in a file based storage (data lake) and/or in a time series database (data warehouse).

II–4.3.2. CADIS data processing

The task of the data processing module is to transform collected measurement data and maintenance reports into the following structured time series:

— Condition history is obtained by applying signal processing in time and frequency domains to the collected data to extract the diagnostics features identified by the component fault matrix.
— Process/context history is obtained by preprocessing (e.g. filtering) the process data (also called contextual data).

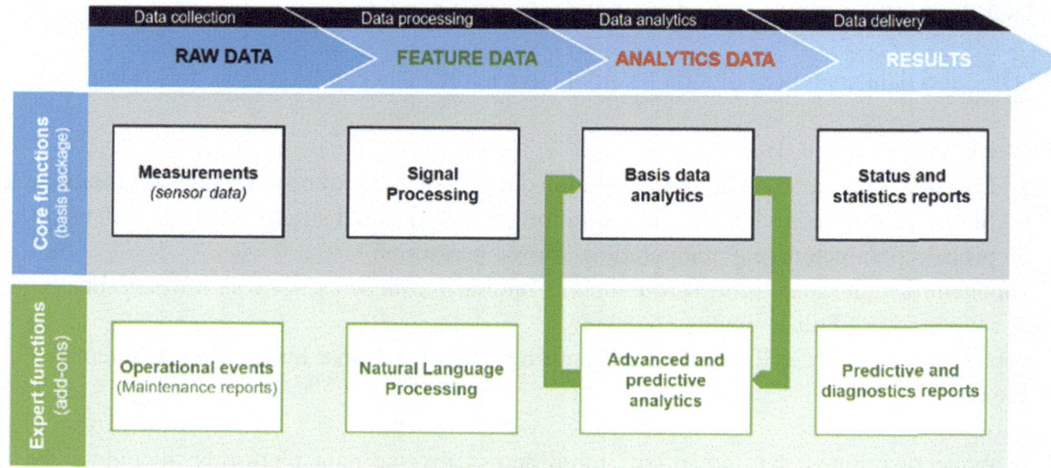

FIG. II–2. CADIS data flow process (courtesy of F. Fomi Wamba, Framatome, Germany).

— Fault history is obtained by extracting fault and health state history from operational logs and maintenance reports using natural language processing.

II–4.3.3. CADIS historian

The task of the historian module is to store the structured time series data (condition history, process history and fault history) within a time series database system. The aim here is to keep the generated information in a format ensuring and/or enabling:

— Storage of the data in various database tables;
— No data redundancy;
— Parallel data retrieval by multiple users or applications;
— Queries based on time intervals (from … to …) and conditions (if …);
— Easy backup and export;
— Data presentation.

II–4.3.4. CADIS data analytics

The task of the data analytics module is to convert the time series into meaningful insights, as shown in Fig. II–3.

(a)   Descriptive analytics on the time series data for anomaly detection and data understanding

Descriptive analytics focuses on understanding time series data and detecting anomalies through univariate analysis, which examines individual parameters, and multivariate analysis, which investigates relationships among multiple parameters.

(i)   Univariate analysis

The purpose of univariate analysis is to describe each time series parameter via a statistical summary in order to find patterns inside the parameter time series data. This method helps to automatically identify the critical diagnostic parameters according to their current status or severity and, finally, to identify the potential component faults according to the correlation matrix component faults versus diagnostic features.

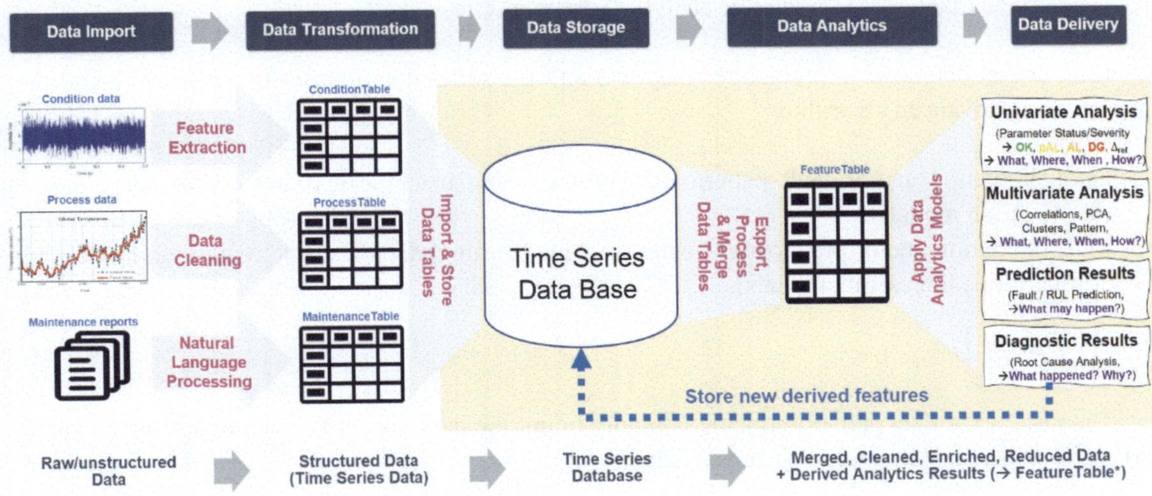

FIG. II–3. Data analytics methods in CADIS (courtesy of F. Fomi Wamba, Framatome, Germany).

Diagnostic parameter status is evaluated by comparing the current parameter value against absolute limits: the preliminary alert limit (pAL), alert limit (AL) and danger limit (DG), for example as derived from standards, from original equipment manufacturer specifications and/or statistics based.

The severity of a diagnostic parameter is quantified as the relative deviation of the current parameter value from the reference value, calculated as shown in Eq. (II–1):

$$\Delta_{ref} = \left| \frac{P_{current} - P_{ref}}{P_{ref}} \right| \times 100 \tag{II–1}$$

where

$P_{current}$ is the current value of a selected parameter;

P_{ref} is the reference value of the selected parameter;

and Δ_{ref} is the relative deviation as a percentage (%).

(ii) Multivariate analysis

The purpose of multivariate analysis is to describe the relationship among several time series parameters via a statistical summary in order to find patterns inside the time series data (or multivariable data distribution):

— Identify correlated/similar features (correlation maps).
— Identify the most important diagnostic features (principal components).
— Identify groups, clusters, outliers and anomalies (clustering, cluster maps).
— Describe identified insights (important features, clusters, correlations).

(b)   Data driven diagnostics (predictive analytics) on time series data

Data driven diagnostics (predictive analytics) on time series data focuses on forecasting and fault detection using advanced algorithms:

— Fault prediction and/or fault pattern recognition (using machine learning and/or deep learning classification models);
— RUL estimation and/or prognostics (using machine learning, deep learning regression models and/or time series forecasting models).

II–4.3.5. CADIS data delivery

The task of the data delivery module is finally to put the data (raw data, feature data and analytics data) at the disposal of engineering teams using:

— Plant asset management and supervision tools (see Fig. II–1);
— Customized dashboards and/or reports;
— Expert tools (for diagnostics, cause analysis, data sciences, analytics model creation);
— Data export interfaces (or data delivery to third party systems).

II–5. CADIS USE CASES

The following use cases illustrate how the CADIS platform applies advanced data analytics to support equipment monitoring and fault diagnosis in nuclear power plants.

II–5.1. Use case 1: Automation of critical asset monitoring

This use case presents the application of CADIS on vibration and process data of a reactor coolant system with main coolant pumps (see Fig. II–4).

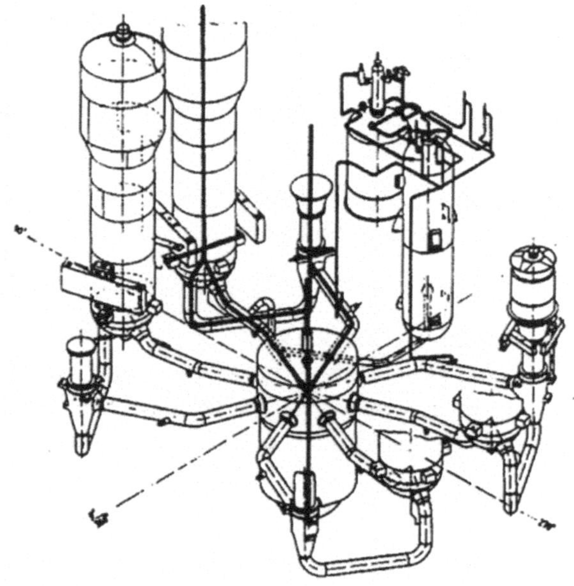

FIG. II–4. Reactor coolant system with main coolant pumps (courtesy of F. Fomi Wamba, Framatome, Germany).

Motor operated equipment, like the reactor coolant pumps in nuclear power plants, is essential for plant operation and power production. Component faults on such equipment can reduce the safety, availability and profitability of the plant and increase the operational cost (e.g. increasing the number of engineering actions).

The measurement data of the reactor coolant system were recorded monthly by the plant operator and provided to Framatome via Internet file transfer. The aim of the Framatome data evaluation is to check the vibrational behaviour of the instrumented components, detect anomalies, identify faults and recommend actions. Table II–2 shows the data flow process automatically performed in CADIS using a configuration file.

The dynamic data were delivered in *.raw format and the static data in *.csv format. These data were imported into the CADIS platform. The data processing applied by CADIS in the time and frequency domains used a configurable moving time window, as defined in Eq. (II–2):

$$\Delta t = \frac{N_S}{F_S} = 32.768 \text{ s} \tag{II–2}$$

where

N_s is the number of samples, defined as $N_s = 2^N$ with $N = 15$;

and F_s is the sampling frequency, equal to 1000 Hz.

For each measurement provided with 900 s record length, approximately 27 non-overlapping time windows (900 s/32.768 s $\approx$ 27) are processed. From each window, diagnostic parameters, such as root mean square (RMS) values, and spectra, such as fast Fourier transform (FFT), power spectral density (PSD) and frequency response function (FRF), are extracted and stored in a time series database. Table II–3 gives an example of the diagnostic parameters and functions extracted during the data processing by CADIS.

TABLE II–2. CADIS DATA FLOW PROCESS

Data import	Data processing	Data storage	Data analytics	Data delivery
Read and import dynamic and static data	For each dynamic channel, time and frequency domain analyses are performed to extract diagnostic features	The diagnostic features (from dynamic data) and the static data are stored in a historian to create structured time series information	Application of the descriptive statistics on time series data (univariate analysis)	Report generation

TABLE II–3. EXAMPLE OF CADIS DIAGNOSTIC PARAMETERS AND FUNCTIONS

Parameter	Description
_A1X	Amplitude 1X
_A2X	Amplitude 2X
_A3X	Amplitude 3X
_A4X	Amplitude 4X
_P1X	Phase 1X
_P2X	Phase 2X
_P3X	Phase 3X
_P4X	Phase 4X
_All	RMS of all frequency lines (0Hz to Fmax), see std or eff
_LF	RMS of all frequency lines below 10 Hz (0 Hz to 10 Hz)
_ISO	RMS of all frequency lines between 10 Hz and 1000 Hz (10 Hz to 1000 Hz)
_HF	RMS of all frequency lines above 1000 Hz (1000 Hz to Fmax)
_Harmonics	Sum of all 10 first Harmonics, see Synchronous
_NonSynchronous	Sum of all Non-Synchronous frequency lines
_SubSynchronous	Sum of all Sub Synchronous frequency lines (below 1X)
_SumAll	Sum of all frequency lines (0 Hz to Fmax)
_Synchronous	Sum of all Synchronous frequency lines (i*1X)
_TH	Sum of all 20 first Harmonics of Frequency Line (50 Hz or 60 Hz)
_RMSWhirl	Whirl energy around 0.41...0.49 * F0
_FreqWhirl	Whirl frequency where Whirl energy = maximum
_mean	Average Value (Mean)
_median	Average Value (Median)
_iqr	Interquartile = Upper Quartil - Lower Quartil
_max	Maximum Value
_min	Minimum Value
_peak	Peak Value = max(abs(Max),abs(Min))

TABLE II–3. EXAMPLE OF CADIS DIAGNOSTIC PARAMETERS AND FUNCTIONS (cont.)

Parameter	Description
_peak2peak	Range = Max - Min
_crest	Crest Factor = Peak/RMS
_kurtosis	Kurtosis
_skewness	Skewness
_eff	RMS, see std or All
_std	Standard Deviation, see eff or All
_histarea	Fläche Histogram
_iqrlowup	iqr upper whisker - iqr lower whisker = whisker range
_nconsts	Number of constant signal sectors sig(ti) = sig(ti-1)
_outliers	Number of outliers
_speed	RotatingSpeed
_irms	integration - filtering - parameter calculation
_ipeak	integration - filtering - parameter calculation
_ipeak2peak	integration - filtering - parameter calculation
_imean	integration - filtering - parameter calculation
_fft_peak_	frequency value in fft spectrum
_fft_mag_	magnitude at the frequency in amplitude spectrum
_fft_ph_	phase at the frequency in amplitude spectrum
_frf_peak_	frequency value in frf spectrum
_frf_mag_	magnitude at the frequency in frf mag spectrum
_frf_ph_	magnitude at the frequency in frf ph spectrum
_frf_coh_	magnitude at the frequency in frf coh spectrum
_psd_peak_	frequency value in psd spectrum
_psd_mag_	magnitude at the frequency in psd spectrum
_fft_rms_	energy value in the frequency range

View of all parameters
of all channels

- Current Status or Severity

- Correlation Parameter vs. Sensor

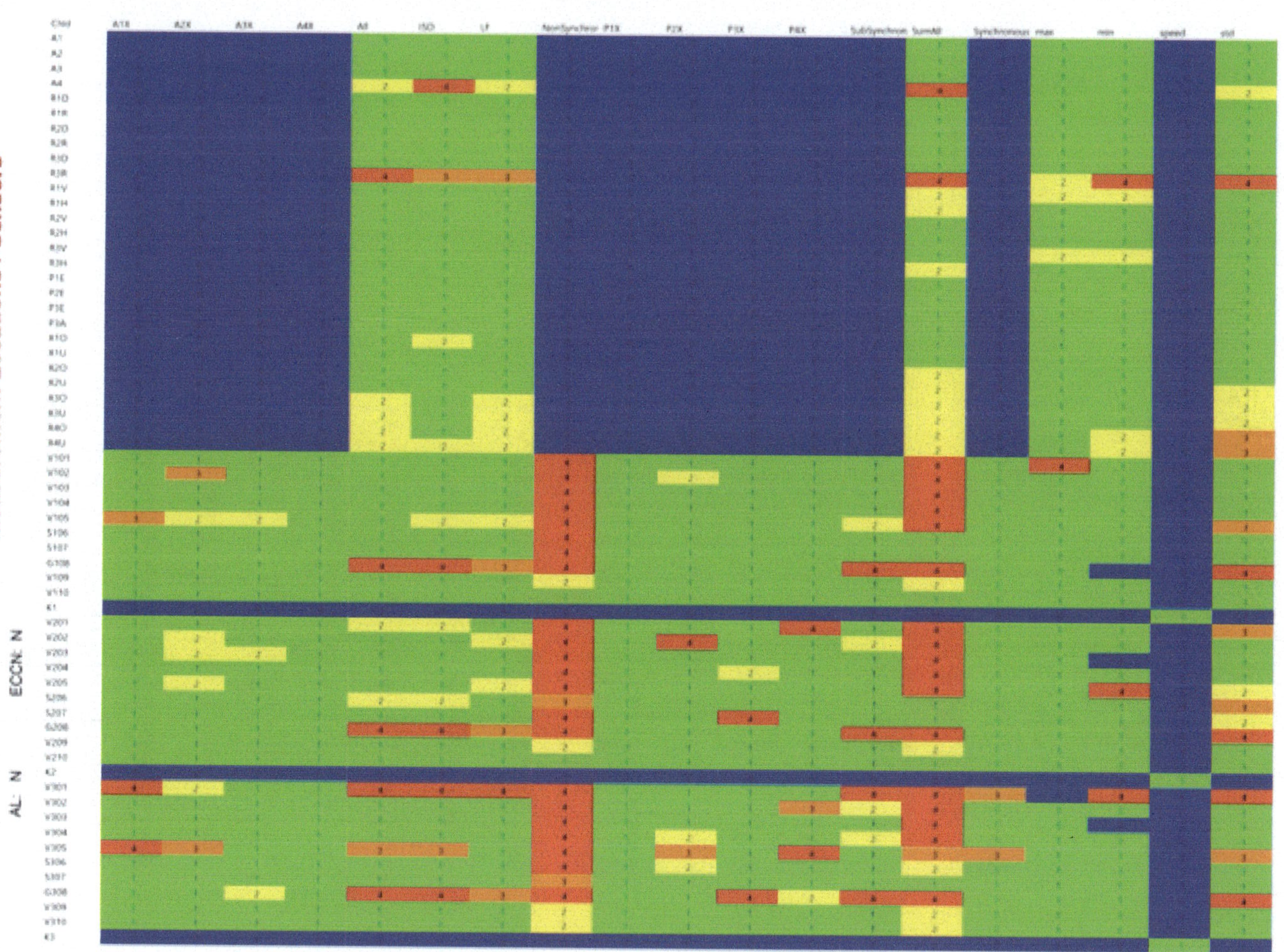

FIG. II–5. Channel parameter status table (courtesy of F. Fomi Wamba, Framatome, Germany).

During this project, descriptive analytics models were applied to the historian data. Predictive analytics was not performed, due to missing fault history data. Figures II–5 to II–9 show some descriptive analytics result examples:

— The channel parameter status table (Fig. II–5) shows the correlation between channels (rows) and parameters (columns) with the current status colour code of the parameters (cells). This table helps identify the critical channels (or signals). The parameter status colour is the result of the univariate descriptive statistics using statistics based limit values. CADIS offers the possibility to manually adapt the statistics based limit values used for anomaly detection.
— The monitoring frequency table (Fig. II–6) shows, for each frequency parameter (peak, magnitude, phase, coherence) of a spectrum (FFT, PSD, FRF), the statistics (Min, Mean, Median, Max, Std, IQR) and status information (CurrentStatus, CurrentValue, DeltaRef, LowerLimit, UpperLimit, RefValue), as well as the potential linked component fault. This enables quick identification of changes in frequency parameters.
— The parameter table (Fig. II–7) shows, for each signal parameter, the statistics and status information, as well as the potential linked component fault. This table helps identify very quickly that parameters and signals need to be investigated in more depth to confirm the existence of a risk for the component.
— The channel info map (Fig. II–8) shows the spectral history, the parameter history, the current/last spectrum and channel parameter status table of a selected channel, sensor or signal. This map supports understanding of the changes and/or anomalies detected in a signal or parameter.
— The orbit info map (Fig. II–9) shows the time waveforms and the spectra of the orbit pair shaft displacement signals, as well as the 1X-filtered and unfiltered orbit plot with the full spectrum (with forward/backward frequencies). This map helps interpret the changes and/or anomalies detected in a shaft displacement signal.

The reports automatically created by CADIS monthly contain detailed information (diagnostic features as time series information and analytics insights such as status, severity and faults) for anomaly detection, anomaly data understanding and diagnosis (cause analysis). These results are useful for optimized asset management, decision making and maintenance recommendations.

During this project, CADIS was able to reduce the time effort by more than 70% and provide better diagnostic information and analytics insights in comparison to previous engineering works (see Fig. II–10). Critical events could be analysed by experts (cause analysis) easily and recommendations issued in a short time.

II–5.2. Use case 2: Cause analysis of a water fluctuation problem

Plant engineers at a nuclear power plant in Europe observed a water level fluctuation problem in a heat exchanger of the secondary circuit in the turbine island over several months. They were unsure how to begin the cause analysis of this problem and asked for assistance from Framatome. Framatome requested all operating and process data from all equipment connected directly or indirectly with the heat exchanger. Historical data for over 1000 different process parameters were delivered, in time periods with problems (fault data) and without problems (health data).

The delivered data were imported into CADIS. After cleaning and time alignment, the data were imported into the CADIS historian. A multivariate analysis (cluster map, see Fig. II–11) was then performed to identify which parameters had the same signature/symptoms as the water level fluctuation (similarity/correlation analysis). After discussion with the plant engineers, a control valve that could not fully open or close when triggered was quickly identified as the cause of the problem.

In this case, CADIS was able to provide the first analytics results and the list of suspected components within a few hours.

Frequency Parameters

ParId	SuesParName	CurrentStatus	CurrentVal	DeltaRef	LowerLimit	RefVal	UpperLimit	MinVal	MeanVal	MedianVal	MaxVal	StdVal	IqrVal	Fault_VibMode
31	A2_A4_frf_mag_5630	0	2.0793	0.0000	0.0000	0.0000	0.0000	0.0123	31.2161	1.4444	1312.1676	141.2807	0.8006	R, v
32	A2_A4_frf_mag_89500	0	0.9985	0.0000	0.0000	0.0000	0.0000	0.0111	1.4579	0.9263	22.1378	2.7964	0.4257	T,v
33	A2_A4_frf_peak_18380	1	18.4326	0.2863	17.5000	18.3800	19.2500	17.8223	18.4688	18.5547	18.9209	0.3617	0.6104	R,v
34	A2_A4_frf_peak_31250	1	31.7383	1.5625	28.5000	31.2500	32.7500	30.6396	31.3659	31.4941	31.7383	0.3737	0.7324	T,v
35	A2_A4_frf_peak_52500	1	52.4902	0.0186	51.0000	52.5000	55.5000	51.8799	52.4132	52.3682	52.9785	0.3253	0.4883	T,v
36	A2_A4_frf_peak_5630	1	5.7373	1.9060	4.8800	5.6300	6.6300	5.1270	5.6292	5.6152	6.1035	0.3185	0.4883	R, v
37	A2_A4_frf_peak_89500	1	89.4775	0.0251	82.5000	89.5000	95.5000	88.9893	89.6392	89.7217	89.9658	0.2890	0.3662	T,v
38	A2_A4_frf_ph_18380	0	169.6586	0.0000	0.0000	0.0000	0.0000	2.0183	166.8453	168.9123	358.4250	60.6406	38.4686	R,v
39	A2_A4_frf_ph_31250	0	4.5689	0.0000	0.0000	0.0000	0.0000	0.0136	150.6495	82.8432	359.9852	150.2955	315.4804	T,v
40	A2_A4_frf_ph_52500	0	352.9068	0.0000	0.0000	0.0000	0.0000	0.0432	217.1113	300.7188	359.9872	147.1020	332.5386	T,v
41	A2_A4_frf_ph_5630	0	190.9148	0.0000	0.0000	0.0000	0.0000	1.0192	178.0798	180.5147	358.3668	59.1778	34.4861	R, v
42	A2_A4_frf_ph_89500	0	5.3139	0.0000	0.0000	0.0000	0.0000	0.6500	124.3439	84.6581	359.5960	103.7962	140.4941	T,v
43	A2_psd_mag_10630	1	0.7496	40.5961	0.1262	1.2618	12.6183	0.0000	1.2392	0.8920	44.7713	2.8128	1.1774	R,v
44	A2_psd_peak_10630	1	10.7117	0.7683	9.5000	10.6300	12.5000	10.1013	10.6277	10.6201	11.1389	0.3047	0.5188	R,v
45	A3_psd_mag_13630	1	4.1052	9.7724	0.4550	4.5499	45.4988	0.0000	6.8661	6.3686	27.3900	3.6085	3.6297	R,v
46	A3_psd_peak_13630	1	13.9465	2.3223	10.6300	13.6300	16.6300	13.1226	13.5614	13.5803	14.1296	0.2096	0.2747	R,v
47	A4_psd_mag_13630	1	2.8011	12.8389	0.3214	3.2137	32.1366	0.0000	2.9263	2.7610	15.1832	2.1467	2.3637	R,v
48	A4_psd_peak_13630	1	13.9465	2.3223	12.6300	13.6300	14.7500	13.1226	13.6809	13.7024	14.1296	0.2130	0.3052	R,v
49	P1E_psd_mag_12630	1	0.0011	0.0000	0.0001	0.0008	0.0076	0.0000	0.0015	0.0014	0.0050	0.0006	0.0007	RC
50	P1E_psd_mag_21380	1	0.0015	0.0000	0.0001	0.0007	0.0075	0.0000	0.0011	0.0011	0.0032	0.0005	0.0006	SW
51	P1E_psd_mag_41250	1	0.0010	0.0000	0.0001	0.0007	0.0070	0.0000	0.0013	0.0013	0.0033	0.0005	0.0006	SW
52	P1E_psd_mag_43130	1	0.0009	0.0000	0.0001	0.0007	0.0068	0.0000	0.0012	0.0012	0.0032	0.0005	0.0006	SW
53	P1E_psd_mag_46000	1	0.0004	0.0000	0.0000	0.0001	0.0014	0.0000	0.0003	0.0003	0.0008	0.0001	0.0001	SW
54	P1E_psd_mag_5500	1	0.0058	0.0000	0.0003	0.0027	0.0275	0.0000	0.0046	0.0044	0.0125	0.0020	0.0023	SW
55	P1E_psd_mag_7630	1	0.0015	0.0000	0.0001	0.0009	0.0087	0.0000	0.0019	0.0018	0.0052	0.0008	0.0009	SW
56	P1E_psd_peak_12630	1	12.2070	3.3489	12.1300	12.6300	13.8800	12.1155	12.6398	12.6343	13.1226	0.2842	0.4578	RC
57	P1E_psd_peak_21380	1	21.6370	1.2019	20.3800	21.3800	22.7500	20.8740	21.3920	21.3928	21.8811	0.2738	0.4272	SW
58	P1E_psd_peak_41250	1	41.0461	0.4942	38.2500	41.2500	42.3800	40.7410	41.2365	41.2292	41.7480	0.2976	0.4883	SW
59	P1E_psd_peak_43130	1	43.3044	0.4045	42.0000	43.1300	46.1300	42.6025	43.1553	43.1824	43.6401	0.2939	0.4883	SW
60	P1E_psd_peak_46000	1	45.6848	0.6852	45.5000	46.0000	46.6300	45.4712	45.8986	45.8679	46.5088	0.3009	0.5188	SW

FIG. II–6. Monitoring frequency table (courtesy of F. Fomi Wamba, Framatome, Germany).

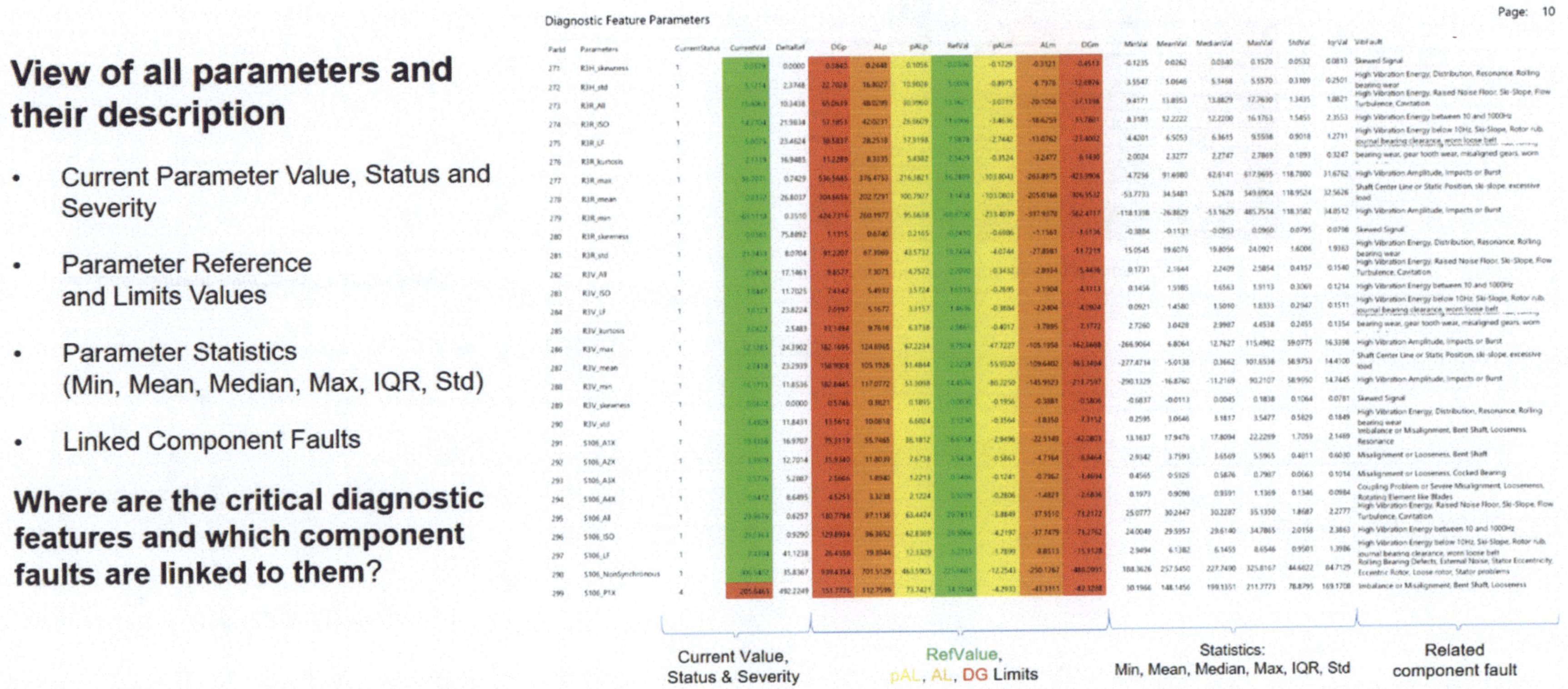

View of all parameters and their description

- Current Parameter Value, Status and Severity

- Parameter Reference and Limits Values

- Parameter Statistics (Min, Mean, Median, Max, IQR, Std)

- Linked Component Faults

Where are the critical diagnostic features and which component faults are linked to them?

Diagnostic Feature Parameters

PartId	Parameters	CurrentStatus	CurrentVal	DeltaRef	DGp	ALp	pALp	RefVal	pALm	ALm	DGm	MinVal	MeanVal	MedianVal	MaxVal	StdVal	IqrVal	VibFault
271	R3H_skewness	1	[illegible]	0.0000	0.3843	0.2048	0.1056	[illegible]	-0.1729	-0.3121	-0.4513	-0.1235	0.0262	0.0340	0.1570	0.0532	0.0813	Skewed Signal
272	R3H_std	1	[illegible]	2.3748	22.7028	16.8027	10.9028	[illegible]	-0.8975	-6.7978	12.6926	3.5547	5.0646	5.3468	5.5570	0.3109	0.2501	High Vibration Energy, Distribution, Resonance, Rolling bearing wear
273	R3R_All	1	[illegible]	10.3438	65.0639	48.0299	30.9960	[illegible]	-3.0719	-20.1058	-17.1198	9.4171	13.8953	13.8829	17.7630	1.3435	1.8821	High Vibration Energy, Raised Noise Floor, Ski-Slope, Flow Turbulence, Cavitation
274	R3R_ISO	1	[illegible]	21.9834	57.1853	42.0237	26.8609	[illegible]	-3.4636	-18.4259	-33.7801	8.3181	12.2222	12.2200	16.1763	1.5455	2.3553	High Vibration Energy between 10 and 1000Hz
275	R3R_LF	1	[illegible]	23.4624	18.3837	28.2518	17.3198	[illegible]	-2.7442	-13.0762	-23.8002	4.4201	6.5053	6.3615	9.5598	0.9018	1.2711	High Vibration Energy below 10Hz, Ski-Slope, Rotor rub, journal bearing clearance, worn loose belt
276	R3R_kurtosis	1	[illegible]	16.9485	11.2289	8.3035	5.4382	[illegible]	-0.3524	-3.2477	-6.1430	2.0024	2.3277	2.2747	2.7869	0.1893	0.3247	bearing wear, gear tooth wear, misaligned gears, worn
277	R3R_max	1	[illegible]	0.7429	536.5685	376.4733	216.3821	[illegible]	-103.8043	-263.8975	-423.9906	4.7256	91.6980	62.6141	617.9695	118.7800	31.6762	High Vibration Amplitude, Impacts or Burst
278	R3R_mean	1	[illegible]	26.8017	304.5656	202.7291	100.7927	[illegible]	-103.0803	-205.0168	306.9532	-53.7733	34.5481	5.2678	549.6904	118.9524	32.5626	Shaft Center Line or Static Position, ski-slope, excessive load
279	R3R_min	1	[illegible]	0.3510	424.7316	260.1977	95.5618	[illegible]	-233.4039	-397.9570	-562.4717	-118.1398	-26.8829	-53.1629	485.7514	118.3582	34.8512	High Vibration Amplitude, Impacts or Burst
280	R3R_skewness	1	[illegible]	75.8892	1.1315	0.6740	0.2165	[illegible]	-0.6696	-1.1563	-1.6136	-0.3884	-0.1131	-0.0953	0.0960	0.0795	0.0798	Skewed Signal
281	R3R_std	3	[illegible]	8.0704	97.2207	67.3069	43.5732	[illegible]	-4.0744	-27.8981	-53.7219	15.0545	19.6076	19.8056	24.0921	1.6006	1.9363	High Vibration Energy, Distribution, Resonance, Rolling bearing wear
282	R3V_All	1	[illegible]	17.1461	9.8577	7.3075	4.7572	[illegible]	-0.3432	-2.8934	-5.4436	0.1731	2.1644	2.2409	2.5854	0.4357	0.1540	High Vibration Energy, Raised Noise Floor, Ski-Slope, Flow Turbulence, Cavitation
283	R3V_ISO	1	[illegible]	11.7025	7.4342	5.4933	3.5724	[illegible]	-0.2695	-2.1904	-4.1113	0.1456	1.5985	1.6563	1.9113	0.3069	0.1214	High Vibration Energy between 10 and 1000Hz
284	R3V_LF	1	[illegible]	23.8224	7.0192	5.1672	3.3157	[illegible]	-0.3884	-2.2404	-4.0904	0.0921	1.4580	1.5010	1.8333	0.2947	0.1511	High Vibration Energy below 10Hz, Ski-Slope, Rotor rub, journal bearing clearance, worn loose belt
285	R3V_kurtosis	1	[illegible]	2.5483	11.1494	9.7616	6.3738	[illegible]	-0.4017	-3.7895	-2.3772	2.7260	3.0428	2.9987	4.4538	0.2455	0.1354	bearing wear, gear tooth wear, misaligned gears, worn
286	R3V_max	1	[illegible]	24.3902	182.1695	124.8065	67.2234	[illegible]	-47.7227	-105.1958	-162.3688	-266.9064	6.8064	12.7627	115.4982	39.0775	16.3398	High Vibration Amplitude, Impacts or Burst
287	R3V_mean	1	[illegible]	23.2939	158.9008	105.1926	51.4844	[illegible]	-55.9320	-109.6402	163.3404	-277.4714	-5.0138	0.3662	101.6538	58.9753	14.4100	Shaft Center Line or Static Position, ski-slope, excessive load
288	R3V_min	1	[illegible]	11.8536	182.8445	117.0772	53.3098	[illegible]	-80.2250	-145.9523	-211.7597	-290.1329	-16.8760	-11.2169	90.2107	58.9950	14.7445	High Vibration Amplitude, Impacts or Burst
289	R3V_skewness	1	[illegible]	0.0000	0.5746	0.3821	0.1895	[illegible]	-0.1956	-0.3881	-0.5806	-0.6837	-0.0113	0.0045	0.1838	0.1064	0.0781	Skewed Signal
290	R3V_std	1	[illegible]	11.8431	13.5612	10.0818	6.6024	[illegible]	-0.3564	-1.8350	-7.3152	0.2595	3.0646	3.1817	3.5477	0.5829	0.1849	High Vibration Energy, Distribution, Resonance, Rolling bearing wear
291	S106_A1X	1	[illegible]	16.9707	75.3119	55.7465	36.1812	[illegible]	-2.9496	-22.5149	-42.0873	13.1637	17.9476	17.8094	22.2259	1.7059	2.1489	Imbalance or Misalignment, Bent Shaft, Looseness, Resonance
292	S106_A2X	1	[illegible]	12.7014	35.9340	11.8039	2.6738	[illegible]	-0.5863	-4.7164	-8.8464	2.9342	3.7593	3.6569	5.5965	0.4811	0.6030	Misalignment or Looseness, Bent Shaft
293	S106_A3X	1	[illegible]	5.2087	2.5666	1.8940	1.2213	[illegible]	-0.1241	-0.7862	-1.4694	0.4565	0.5926	0.5876	0.7987	0.0663	0.1014	Misalignment or Looseness, Cocked Bearing
294	S106_A4X	1	[illegible]	8.6495	4.5251	3.3238	2.1224	[illegible]	-0.2806	-1.4821	-2.5836	0.1973	0.9098	0.9391	1.1369	0.1346	0.0984	Coupling Problem or Severe Misalignment, Looseness, Rotating Element like Blades
295	S106_All	1	[illegible]	0.6257	180.7798	97.1136	63.4424	[illegible]	-3.8849	-37.5510	-71.2172	25.0777	30.2447	30.2287	35.1350	1.8687	2.2777	High Vibration Energy, Raised Noise Floor, Ski-Slope, Flow Turbulence, Cavitation
296	S106_ISO	1	[illegible]	0.9290	129.8934	36.3652	62.8369	[illegible]	-4.2197	-37.7479	-71.2762	24.0049	29.5957	29.6140	34.7865	2.0158	2.3863	High Vibration Energy between 10 and 1000Hz
297	S106_LF	1	[illegible]	41.1238	26.4558	19.3544	12.3329	[illegible]	-1.7899	-8.8513	-15.9128	2.9494	6.1382	6.1455	8.6546	0.9501	1.3986	High Vibration Energy below 10Hz, Ski-Slope, Rotor rub, journal bearing clearance, worn loose belt
298	S106_NonSynchronous	1	[illegible]	35.8367	939.4354	701.5129	463.5905	[illegible]	-12.2543	-250.1767	-488.0991	188.3626	257.5450	227.7490	325.8167	44.6022	84.7129	Rolling Bearing Defects, External Noise, Stator Eccentricity, Eccentric Rotor, Loose rotor, Stator problems
299	S106_P1X	4	205.6465	492.2249	151.7726	112.7599	73.7421	14.7594	-4.2933	-43.3111	-42.3208	30.1966	148.1456	199.1351	211.7773	78.8795	169.1708	Imbalance or Misalignment, Bent Shaft, Looseness

Current Value, Status & Severity — RefValue, pAL, AL, DG Limits — Statistics: Min, Mean, Median, Max, IQR, Std — Related component fault

FIG. II–7. Parameter table (courtesy of F. Fomi Wamba, Framatome, Germany).

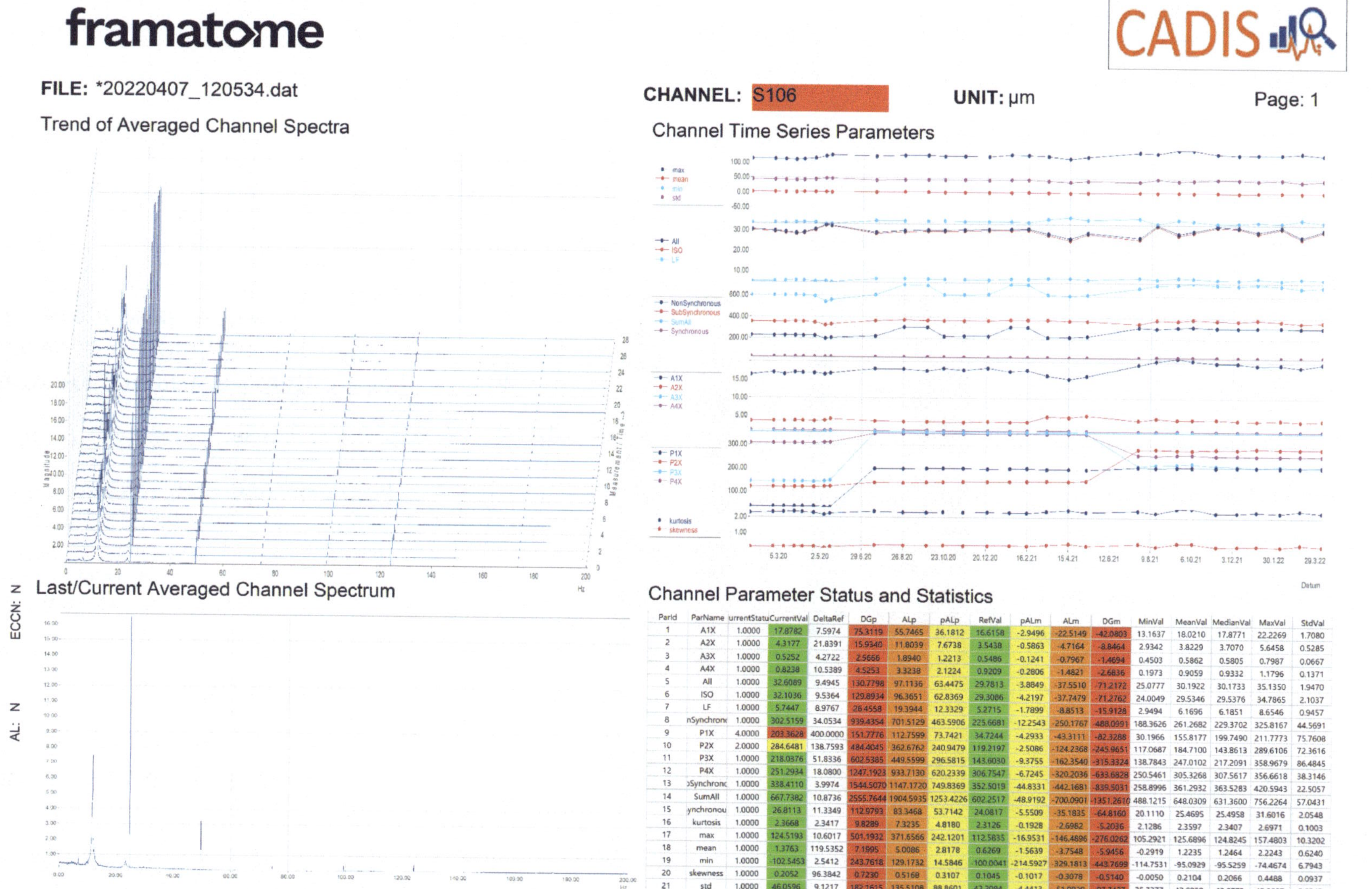

ParId	ParName	CurrentStatus	CurrentVal	DeltaRef	DGp	ALp	pALp	RefVal	pALm	ALm	DGm	MinVal	MeanVal	MedianVal	MaxVal	StdVal
1	A1X	1.0000	17.8782	7.5974	75.3119	55.7465	36.1812	16.6158	-2.9496	-22.5149	-42.0803	13.1637	18.0210	17.8771	22.2269	1.7080
2	A2X	1.0000	4.3177	21.8391	15.9340	11.8039	7.6738	3.5438	-0.5863	-4.7164	-8.8464	2.9342	3.8229	3.7070	5.6458	0.5285
3	A3X	1.0000	0.5252	4.2722	2.5666	1.8940	1.2213	0.5486	-0.1241	-0.7967	-1.4694	0.4503	0.5862	0.5805	0.7987	0.0667
4	A4X	1.0000	0.8238	10.5389	4.5253	3.3238	2.1224	0.9209	-0.2806	-1.4821	-2.6836	0.1973	0.9059	0.9332	1.1796	0.1371
5	All	1.0000	32.6089	9.4945	130.7798	97.1136	63.4475	29.7813	-3.8849	-37.5510	-71.2172	25.0777	30.1922	30.1733	35.1350	1.9470
6	ISO	1.0000	32.1036	9.5364	129.8934	96.3651	62.8369	29.3086	-4.2197	-37.7479	-71.2762	24.0049	29.5346	29.5376	34.7865	2.1037
7	LF	1.0000	5.7447	8.9767	26.4558	19.3944	12.3329	5.2715	-1.7899	-8.8513	-15.9128	2.9494	6.1696	6.1851	8.6546	0.9457
8	nSynchrone	1.0000	302.5159	34.0534	939.4354	701.5129	463.5906	225.6681	-12.2543	-250.1767	-488.0991	188.3626	261.2682	229.3702	325.8167	44.5691
9	P1X	4.0000	203.3628	400.0000	151.7776	112.7599	73.7421	34.7244	-4.2933	-43.3111	-82.3288	30.1966	155.8177	199.7490	211.7773	75.7608
10	P2X	2.0000	284.6481	138.7593	484.4045	362.6762	240.9479	119.2197	-2.5086	-124.2368	-245.9651	117.0687	184.7100	143.8613	289.6106	72.3616
11	P3X	1.0000	218.0376	51.8336	602.5385	449.5599	296.5815	143.6030	-9.3755	-162.3540	-315.3324	138.7843	247.0102	217.2091	358.9679	86.4845
12	P4X	1.0000	251.2934	18.0800	1247.1923	933.7130	620.2339	306.7547	-6.7245	-320.2036	-633.6828	250.5461	305.3268	307.5617	356.6618	38.3146
13	Synchrone	1.0000	338.4110	3.9974	1544.5070	1147.1720	749.8369	352.5019	-44.8331	-442.1681	-839.5031	258.8996	361.2932	363.5283	420.5943	22.5057
14	SumAll	1.0000	667.7382	10.8736	2555.7644	1904.5935	1253.4226	602.2517	-48.9192	-700.0901	-1351.2610	488.1215	648.0309	631.3600	756.2264	57.0431
15	ynchronou	1.0000	26.8113	11.3349	112.9793	83.3468	53.7142	24.0817	-5.5509	-35.1835	-64.8160	20.1110	25.4695	25.4958	31.6016	2.0548
16	kurtosis	1.0000	2.3668	2.3417	9.8289	7.3235	4.8180	2.3126	-0.1928	-2.6982	-5.2036	2.1286	2.3597	2.3407	2.6971	0.1003
17	max	1.0000	124.5193	10.6017	501.1932	371.6566	242.1201	112.5835	-16.9531	-146.4896	-276.0262	105.2921	125.6896	124.8245	157.4803	10.3202
18	mean	1.0000	1.3763	119.5352	7.1995	5.0086	2.8178	0.6269	-1.5639	-3.7548	-5.9456	-0.2919	1.2235	1.2464	2.2243	0.6240
19	min	1.0000	-102.5453	2.5412	243.7618	129.1732	14.5846	-100.0041	-214.5927	-329.1813	-443.7699	-114.7531	-95.0929	-95.5259	-74.4674	6.7943
20	skewness	1.0000	0.2052	96.3842	0.7230	0.5168	0.3107	0.1045	-0.1017	-0.3078	-0.5140	-0.0050	0.2104	0.2066	0.4488	0.0937
21	std	1.0000	46.0596	9.1217	182.1615	135.5108	88.8601	42.2094	-4.4413	-51.0920	-97.7427	35.7277	42.6959	42.6778	49.2657	2.6943

FIG. II–8. Channel info map (courtesy of F. Fomi Wamba, Framatome, Germany).

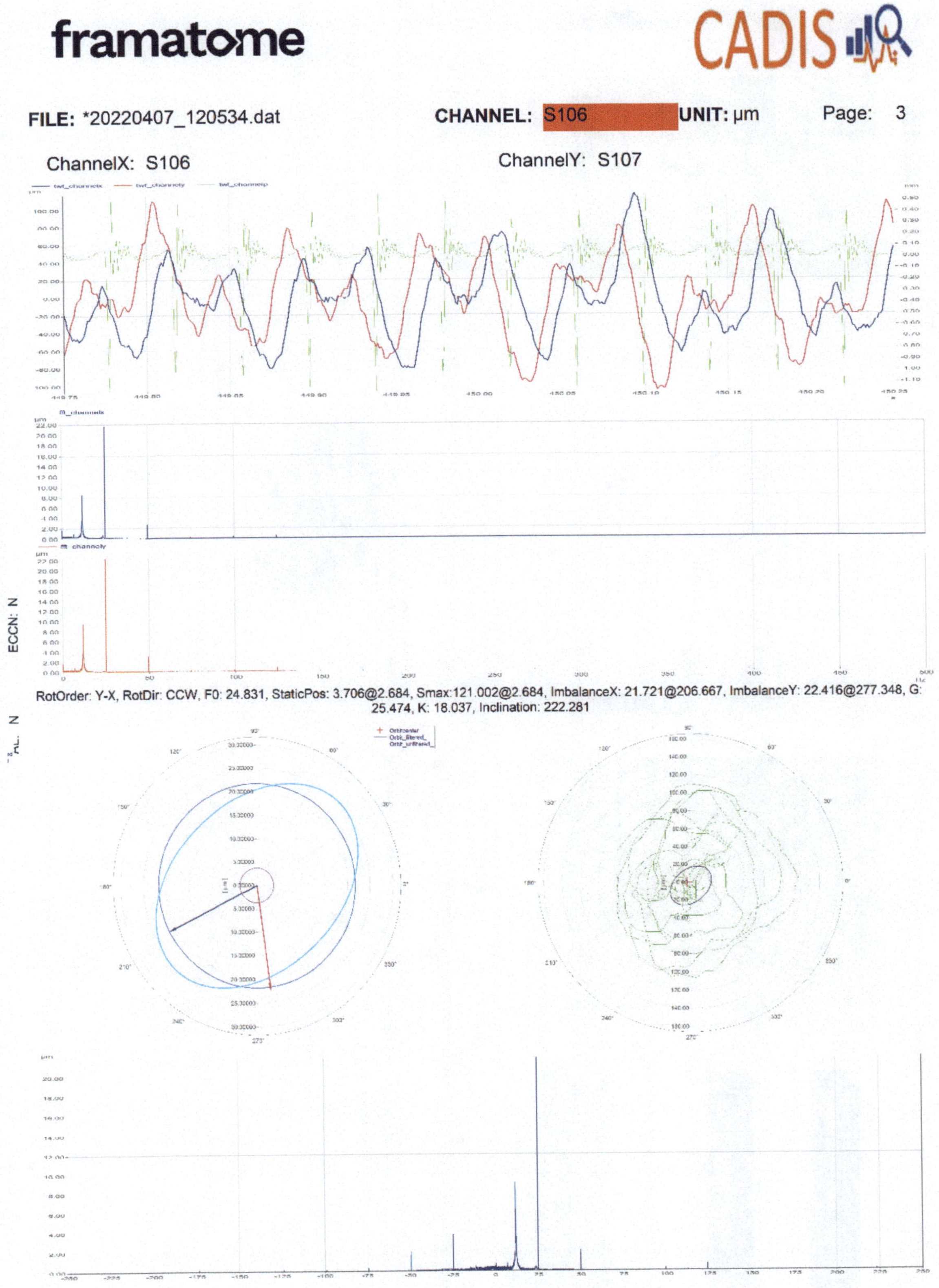

FIG. II–9. Orbit info map (courtesy of F. Fomi Wamba, Framatome, Germany).

113

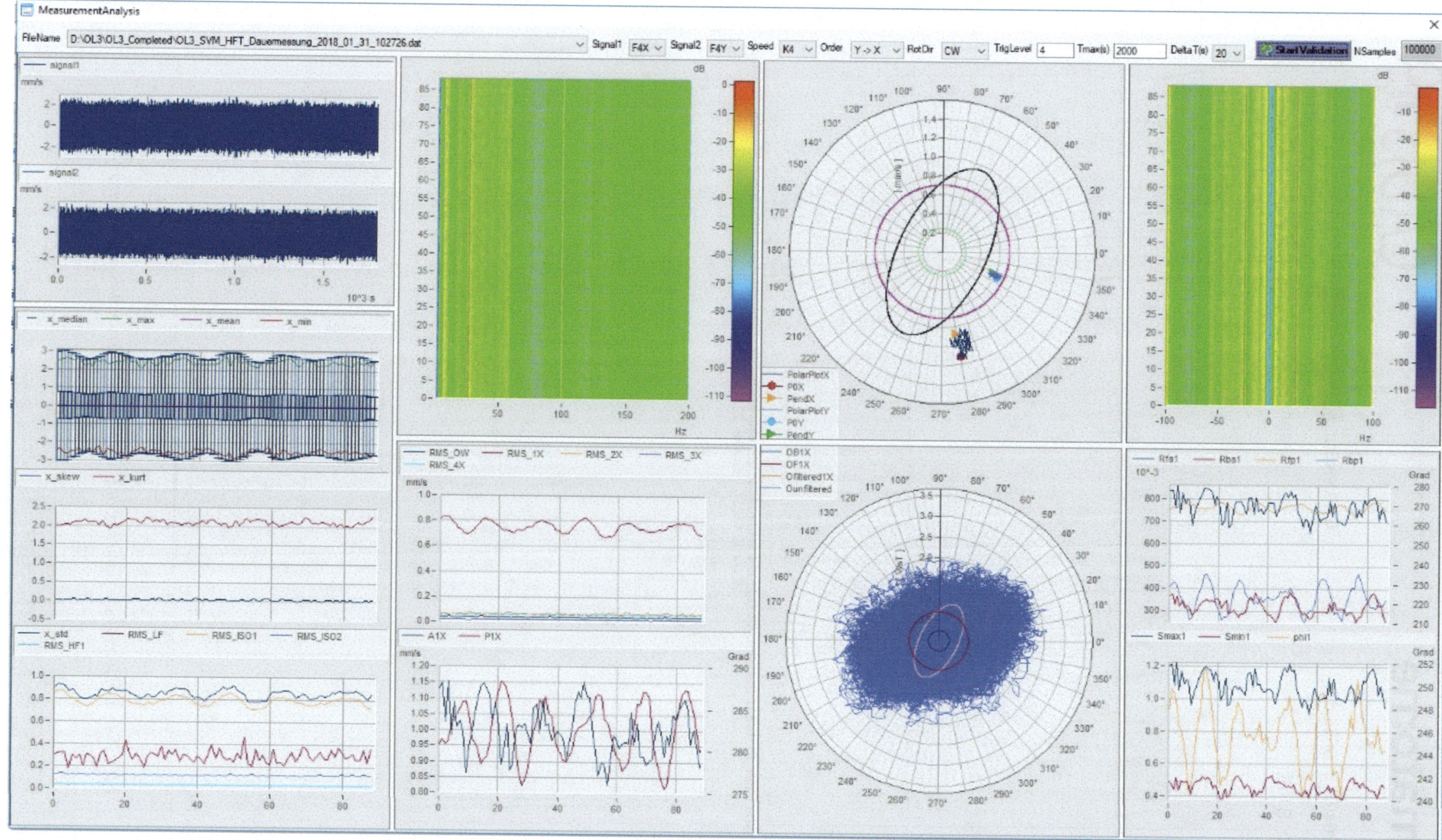

FIG. II–10. CADIS expert tool: vibration analysis for cause analysis (courtesy of F. Fomi Wamba, Framatome, Germany).

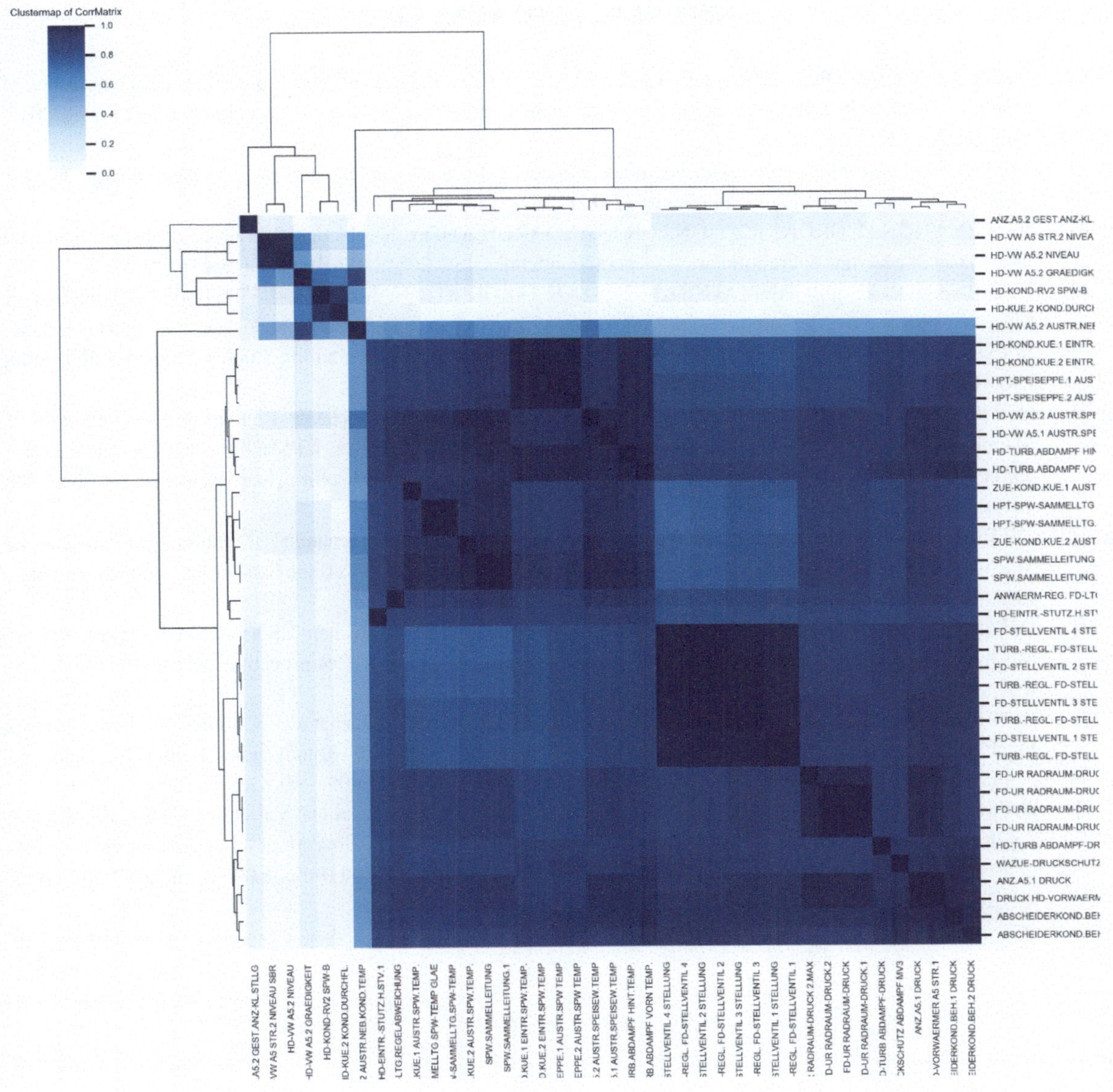

FIG. II–11. Multivariate analysis (cluster map) results using CADIS (courtesy of F. Fomi Wamba, Framatome, Germany).

II–6. CONCLUSION

CADIS automates data collection and analyses for maintenance planning and decision support and reduces engineering costs by optimizing the safety and availability of instrumented assets.

CADIS has already been deployed in several projects and applications as an off-line data analytics and decision support tool. The RPA and AI methods used in CADIS, together with industry standard communication interfaces, signal/text/natural language processing, anomaly detection, fault pattern recognition, statistics based RUL estimation, machine learning and deep learning, are mature, proven techniques. Therefore, all new nuclear power plants designed or modernized by Framatome are systematically equipped with the CADIS solution.

REFERENCES TO ANNEX II

[II–1]   INTERNATIONAL ORGANIZATION FOR STANDARDIZATION, Condition Monitoring and Diagnostics of Machines — Data Interpretation and Diagnostics Techniques — Part 1: General Guidelines, ISO 13379-1:2012, ISO, Geneva (2012).

[II–2]   THOMAS, K., Reactor analytics drives nuclear industry towards machine learning, Nuclear Energy Insider, Analysis for the nuclear energy community (2017).

[II–3]   INTERNATIONAL ORGANIZATION FOR STANDARDIZATION, Mechanical Vibration — Measurement and Evaluation of Machine Vibration Part 1: General Guidelines, ISO 20816-1:2016 ISO, Geneva (2016).

[II–4]   INTERNATIONAL ORGANIZATION FOR STANDARDIZATION, Mechanical Vibration — Evaluation of Machine Vibration by Measurements on Non-rotating Parts — Part 2: Land-based Steam Turbines and Generators in Excess of 50 MW with Normal Operating Speeds of 1 500 r/min, 1 800 r/min, 3 000 r/min and 3 600 r/min, ISO 10816-2:2009, ISO, Geneva (2009).

[II–5]   INTERNATIONAL ORGANIZATION FOR STANDARDIZATION, Mechanical Vibration — Evaluation of Machine Vibration by Measurements on Non-rotating Parts — Part 3: Industrial Machines with Nominal Power above 15 kW and Nominal Speeds between 120 r/min and 15 000 r/min when Measured in Situ, ISO 10816-3:2009, ISO, Geneva (2009).

[II–6]   INTERNATIONAL ORGANIZATION FOR STANDARDIZATION, Mechanical Vibration — Evaluation of Machine Vibration by Measurements on Non-rotating Parts — Part 4: Gas Turbine Sets with Fluid-film Bearings, ISO 10816-4:2009, ISO, Geneva (2009).

[II–7]   INTERNATIONAL ORGANIZATION FOR STANDARDIZATION, Mechanical Vibration — Evaluation of Machine Vibration by Measurements on Non-rotating Parts — Part 6: Reciprocating Machines with Power Ratings above 100 kW, ISO 10816-6:1995, ISO, Geneva (1995).

[II–8]   INTERNATIONAL ORGANIZATION FOR STANDARDIZATION, Mechanical Vibration — Evaluation of Machine Vibration by Measurements on Non-rotating Parts — Part 7: Rotodynamic Pumps for Industrial Applications, Including Measurements on Rotating Shafts, ISO 10816-7:2009, ISO, Geneva (2009).

[II–9]   INTERNATIONAL ORGANIZATION FOR STANDARDIZATION, Mechanical Vibration — Measurement and Evaluation of Machine Vibration — Part 2: Land-based Gas Turbines, Steam Turbines and Generators in Excess of 40 MW with Fluid-film Bearings and Rated Speeds of 1 500 r/min, 1 800 r/min, 3 000 r/min and 3 600 r/min, ISO 20816-2:2017, ISO, Geneva (2017).

FLEXIBLE OPERATION SUCCESS STORY AT KWU NUCLEAR POWER PLANT WITH VARIOUS DIGITAL TECHNOLOGIES

In the 1980s, Siemens Kraftwerk Union (KWU), the predecessor company of today's Framatome, introduced enhanced capability for flexible operation (FlexOp) into the original design of both pressurized water reactors (PWRs) and boiling water reactors (BWRs). For PWRs, the design characteristics included:

— A part load diagram with constant average coolant temperature in the upper load region, reducing the impact of load changes on primary circuit components;
— Optimized management and treatment of the primary coolant;
— A control rod manoeuvring programme enabling precise control of the axial power distribution;
— Accurate in-core measurement of the power distribution, mostly in real time.

In addition, manoeuvring guidelines were developed with the aim of achieving greater operating margins for FlexOp and preventing pellet–cladding interactions.

For BWRs, there were similar design characteristics, plus recirculation control and manoeuvring guidelines [III–1, III–2].

Initially, these capabilities were used only occasionally to address grid related events, and the plants were mainly optimized for base load operation. Over the past two decades, with the rapidly increasing but fluctuating power generated by renewable energy sources, a strong need for frequent FlexOp appeared for nuclear power plants. Since 2008, the German electricity market has allowed negative electricity prices, and the participation in additional ancillary and balancing markets became attractive due to high price levels. As a result, electrical utilities started to increase stepwise the ranges for the various grid services to be performed by nuclear power plants. KWU-build plants are able to perform both primary and secondary frequency control, as well as minute/tertiary reserve and 'classical' load following Ref. [III–2]. Advanced instrumentation and control (I&C) systems have been introduced to help operators perform transient changes in load and to maximize the level of potential and actual flexibility. Extended low power operation also requires evaluation and optimization. It has been carried out with respect to appropriate fuel conditioning [III–3], reactivity management and mitigation of ageing mechanisms such as flow induced corrosion (FIC), fatigue and vibration. As an overall result, a global improvement of plant performance has been realized, in particular with optimized operation envelopes, inspection intervals and maintenance plans.

This annex describes the advanced digital technologies that have helped to realize this global improvement.

III–1. ADVANCED INSTRUMENTATION AND CONTROL TO SUPPORT OPERATORS DURING MANOEUVRES

The first FlexOp projects focused on improving turbine I&C systems (e.g. at the PWRs at Neckarwestheim II and Biblis units A and B in Germany and Gösgen in Switzerland), followed by improvements of core control. The highest levels of flexibility within the given limits in all phases of the burnup cycle were achieved by the PWRs that introduced advanced load following control (ALFC) as an upgrade of the TELEPERM-XS reactor control system. This covers xenon (Xe) calculation, automated reactivity management and optimized control of axial power distribution that inhibits any axial oscillation of the axial power distribution. It also allows, in case of stochastic load changes, the performance of FlexOp without manual intervention [III–4].

ALFC was further improved by adding predictor technology, including a visual display of the parameters that are controlled for management of the reactivity [III–5]. Overall, the impact of FlexOp on systems and components was minimized. To date, five ALFC projects have been realized in KWU-build PWRs in Germany and Switzerland. Framatome is currently further developing predictor technology for FlexOp as a separate expert system for different types of plant (e.g. see Ref. [III–6]).

ALFC does not affect defence in depth: I&C safety functions have a higher priority and bring the plant to a safe state even in the case of ALFC failure.

III–1.1. Load following features

A special I&C function, called the Load Governor (see Ref. [III–7]), can be used to automate load following. In 2002, Framatome first installed it in Neckarwestheim II. It enables an automatic management of electrical power according to the load schedule issued by the grid dispatcher.

The Load Governor automatically changes the plant output according to the load schedule by acting on the power set point of the turbine controller. It reduces the effort for the turbine operator considerably. It can also take over the functions of primary and secondary frequency control if the existing turbine controller does not support them.

Mitigation of the possible negative effects of FlexOp on the reactor core can be supported by a special I&C function called the Core Governor [III–6, III–8], which may be deployed in the future. Its task is to keep the reactor core away from the limits of limitation and protection systems and avoid Xe oscillations during FlexOp.

III–1.2. Frequency control features for a conventional island

The electrical output of a PWR is controlled by the positioning of the turbine inlet valves. Primary and secondary controls are provided by the turbine controller.

— Primary frequency control: The turbine controller gets the actual value of the grid frequency by measuring the turbine speed. A special factor called 'static', multiplied by the deviation from the nominal grid frequency (50 Hz in Europe) gives the required power correction. The turbine controller adjusts the turbine inlet valves to achieve this correction.
— Remote secondary control: The grid dispatcher determines the target set point of the generator output of the turbine controller, with the maximum ramp rate and power range having been agreed upon beforehand. The actual activation is carried out by the plant shift on request of the grid dispatcher, resulting from market auction and transmission system operator (grid operator) requirements.

Generally, the primary frequency control is combined with the other existing flexible operating modes. In several German and Swiss PWRs, the first step to fulfil the increased grid requirements for plant flexibility was reached by digital upgrades and/or improvements of turbine controllers. Fulfilment of grid requirements before and after optimization of the turbine control is shown in Fig. III–1.

III–1.3. Power control features for a nuclear island

Implementation of ALFC enables PWR operators to perform various non-baseload operation modes in a precise manner, meeting grid requirements and increasing the range of the services provided. Higher ramp rates could be achieved by the plant operators relying on automatic approaches that are well trained (on a simulator) and well practised.

Reference [III–4] shows ALFC handling the axial power density distribution during a commissioning test in 2014 for the Isar 2 PWR, with each frame showing an axial power distribution profile based on

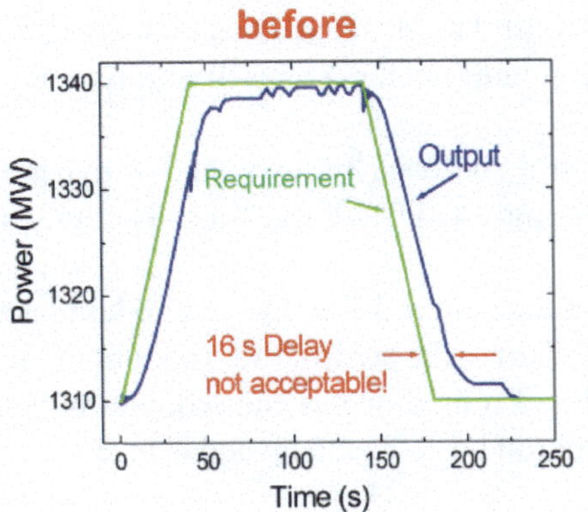

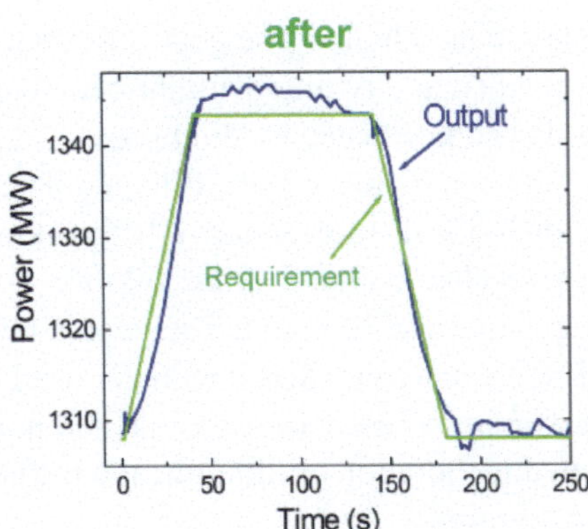

FIG. III–1. Fulfilment of grid requirements before and after turbine control optimization (reproduced from Ref. [III–7] with permission).

in-core signals. It can be seen that ALFC is solving the important task of keeping the profile in equilibrium by damping of the axial Xe oscillations during the whole cycle. Even at the most challenging moment, when the plant reached full load — 100% of the rated electrical output (REO) — and the distance to the boundaries of power limitation functions (orange bars) was the smallest, no manual actions were required. Appropriate modelling of safety limits using sufficiently simplified models and filtering of the in-core signals, both inside the limitation system, made a valuable contribution to the resulting performance.

Digital I&C has been used by ALFC to improve process technology with regard to enhanced load flexibility with respect to:

— Adaptation to shifting reactivity coefficients for transparent automatic reactivity management by reactivity balancing and for better adaptation to changing core loads — low leakage, increased enrichment, mixed oxide fuel (MOX), etc. — opening possibilities for power uprates;
— Self-adaptation to changes in power distribution that are dependent on fuel burnup;
— Exact mixing of boric acid and demineralized water, especially at the end of the fuel cycle with very low boron concentration in the coolant;
— Reactivity management, including on-line Xe calculation for the power distribution controller and management of the boric acid and demineralized water injection;
— Adaptive filtering to reduce signal noise (especially fluctuation of neutron flux) and to allow operation within smaller margins.

Based on the collected operational experience, ALFC was further developed, in cooperation with the utility, to include predictive capabilities [III–4]. Reference [III–5] shows the reactivity balance just before a 30 MW(e)/min up-ramp is initiated after a 6 hour period at low power, during a predictor commissioning test "100% – 47% – 100% REO". The balance includes, among other factors, the predicted Xe contribution during the up-ramp. The arrows show the decomposition of the overall reactivity to single process values and actuators, such as the control rods or the chemical volume control system (CVCS). It shows a reserve or need of reactivity depending on the arrow direction related to the power increase back to the full load. Among other advantages, this visualization allows operators to comply with the World Association of Nuclear Operators requirement for the plausible reactivity management during FlexOp in a suitable and automated way.

In 2021, reactivity management within ALFC was further improved with regards to extended low power operation, which was optimized and fully automated, supporting different strategies depending on the planned duration at reduced power level.

— Duration less than 8 hours: The goal is to keep the control rod bank at a position where full load can be safely reached at any time with the required ramp rate (i.e. the Xe reactivity loss during the part load period is compensated by dilution).
— Duration between 8 hours and 30 hours: The control rod bank follows the Xe dynamics, saving boron/demineralized water. Shortly before the up-ramp, ALFC adjusts bank reactivity so that full load can be reached without further dilution.
— Duration greater than 30 hours: The goal is to keep the core rod-free for a more homogeneous fuel burnup. The Xe reactivity loss is initially used to withdraw the control rod bank out of the core until the Xe maximum is reached. After that and until the waiting time has elapsed, the Xe dynamics are compensated by boration or dilution, and the control rod bank is held up and is inserted only shortly before the up-ramp.

These strategies were extensively validated using a plant simulator and successfully approved during the commissioning phase. The first strategy is already receiving positive operation feedback.

In the future, prediction and optimization strategies can be adapted for different nuclear power plants, enabling the possibility to save boron and continuously provide the power requested without manual interventions. The biggest advantages are expected with fleet management.

III–2. AGEING CONSIDERATIONS

Digital solutions play an important role in evaluating and mitigating the possible ageing impacts of FlexOp. Overall, operational experience in KWU plants shows that FlexOp can be performed safely and reliably and that ageing impacts can be mitigated by developing suitable operational envelopes, by analysis of components affected by transients, by applying detailed advanced methodologies and by appropriate monitoring. The combination of monitoring with centralization of operational data can bring various advantages. Monitoring and maintenance can be continuously adapted to the planned flexibility level of the nuclear power plant and reduce operational costs.

III–2.1. Fatigue aspects

Fatigue is a central concern in nuclear power plant FlexOp. The fatigue status of nuclear power plants can be monitored in several ways. The simplest approach is counting the actual load cycles/transients during lifetime compared against the total cycle number specified in the so-called load specification for design transients.

As the German nuclear power plants were mostly operated in base load mode until 2008, there was considerable reserve for cycling within FlexOp.

In addition to periodic non-destructive testing, particularly for safety related components, preventive measures were taken to diminish the impacts of FlexOp on material fatigue, such as the optimization of operational control and more realistic fatigue analyses. The replacement of components at fatigue relevant locations (e.g. tee pieces) is only a last resort measure.

To support the challenges of nuclear power plant long term operation, including the introduction of FlexOp, Framatome developed the Advanced Fatigue Solution (AFS) (see Ref. [III–9]). In the case of reduced available margins, the existing conservatism needs to be reduced by introducing more realistic input for thermal load assumptions and/or more detailed methodology. AFS addresses both load determination (including load monitoring) and fatigue assessment:

— Load determination:
  • The thermal load assumptions in design transients are the main source of conservatism. That is why the determination of the local thermal loads (temperature differences and gradients) are a key issue. Framatome relies on the determination of the local thermal loads by application

of temperature measurement sections, provided by the load monitoring system FAMOSi hardware, at fatigue relevant locations (e.g. seven thermocouples over the pipe circumference close to the feedwater nozzle).

- If local measurements are not available or not in a suitable position (e.g. in the case of the surge line), then hydraulic transfer functions derived by computational fluid dynamics simulation, for example, can be applied.
- The locally determined loads can be used for the update of design transients of the load specification to (less conservative) model transients or taken as a direct input to more realistic fatigue calculations.
- Framatome has developed the tool TraCo for detection of transients. It can be used for the automatic update of model transients and for delivering the number of cycles (automatic transient counting).

— Fatigue assessment modules with increasing effort and decreasing conservatism:

- Simplified fatigue estimation (SFE) delivers an initial conservative estimate of fatigue relevance via simplified evaluation of the corresponding cumulative usage factor (CUF). SFE is widely applied and is based on local measurements provided by fatigue monitoring systems. It has already resulted in additional margins and operational optimizations.
- Fast fatigue evaluation (FFE) delivers CUF according to all requirements of the design code (e.g. including the consideration of environmentally assisted fatigue (EAF) as a new requirement). As an example, FFE was applied to a flange and surge line of different PWRs based on available local measurements to reduce conservatism [III–10, III–11].
- Detailed fatigue calculation (DFC) is based on the model transients for critical components with respect to FlexOp (non-linear cyclic material behaviour can be considered in the sense of further margin reduction). In German nuclear power plants, DFC was applied if the CUF needed further reduction (e.g. the spray line for PWRs [III–12]).

This graded approach is highly beneficial for optimized plant operation and is dependent on the complexity of the task to be fulfilled, saving time and cost for the realization. In this way, it is possible to recover fatigue margins for new requirements, such as increased FlexOp and the consideration of EAF in the framework of tightening design code requirements in an international context.

More recently, the fatigue related phenomena of vibrational fatigue (VF) and high cycle fatigue (HCF) became relevant development topics, particularly in terms of related fatigue monitoring capabilities:

— The VF monitoring system FAMOS-V allows for keeping VF (e.g. small bore piping, lean tower structures in conventional plants) under control using few accelerometer measurements for load determination [III–13]. The first successful applications exist for conventional plants and would be beneficial for the nuclear field.

— HCF is characterized by irregular temperature fluctuations of small amplitudes and high numbers of cycles. For example, it occurs in mixing zones (e.g. leakage in the heat removal system [III–14]) or in terms of vortex penetration in non-isolable branch lines (e.g. leakage in the safety injection lines and drain lines [III–15]). Framatome implemented local temperature measurement sections and additional draw wire instrumentation for (local) dilation measurement in different PWRs on HCF relevant locations. Such sensors allow for the collection of additional information on the integral dilation of pipe sections due to temperature fluctuations and due to special HCF load effects. Furthermore, a screening method for the identification of the relevant locations was developed and successfully applied [III–16].

These new monitoring capabilities allow for an appropriate management of VF and HCF effects, avoiding additional outage time and even replacement of components.

III–2.2. Predictive maintenance

To study the influence of FlexOp on PWR performance and reliability, a data analysis solution named CADIS (see also Annex II) was applied. It was developed by Framatome for automated data driven diagnostics with the aim of supporting the knowledge based or expert based diagnostics using component data, such as condition, process and log data [III–17]. As the result of such application, recommendations were made for a deeper evaluation of the influence of FlexOp. Several components of the secondary circuit that were expected to be most impacted were chosen. As the next step, the pilot phase was defined to show the applicability of the methodology by creating a dedicated test monitoring system on five components. Overall, information from vibration and more than 1100 operational signals was combined.

One of the observed results of these tests has been the highly varying vibration levels on the condensate pipe during power increase (see Fig. III–2 [III–18]). The upper curve shows the corresponding generator active power increasing from zero to full power over about 2 hours. The lower curve shows the root mean square values of the acceleration signal at various measurement positions over the time. For example, the green curve starts at a fairly constant low level and develops to oscillating level between 0.14 and 0.35 g effective values at full power.

Subsequently, a dedicated detailed finite element method (FEM) stress and fatigue analysis was performed on that pipe, with a calculation of related CUF to verify when the replacement or maintenance should be planned.

With the CADIS tool, multivariate statistical analyses like correlation and cluster map analysis were performed on the vibration and operational data to highlight similarities on separated system parameters and to better understand the cause–effect relationship and the fault signature propagation.

Overall, the pilot showed the applicability of the methodology and, in the case of the advanced FlexOp, it can be deployed for other components of the secondary circuit (about 40 components have been identified). The main advantages of such a methodology are stated to be improved asset management and plant safety, as well as additional cost saving options with condition based/predictive maintenance.

III–2.3. Flow induced corrosion

One of the important degradation processes to be analysed with respect to planned FlexOp is related to FIC. FIC — also known as flow accelerated corrosion (FAC), liquid droplet impingement corrosion and cavitation corrosion — is a degradation process resulting in wall thinning of piping, vessels, heat exchangers and other equipment made of carbon and low alloy steel. The FIC degradation mechanism occurs only locally, under specific conditions of flow, water chemistry, temperature and materials applied [III–19], and it is generally known as one of the most frequently experienced causes of component failures for the plant [III–20].

For operation of the plant in part load conditions, throttling of flow rates may be needed at valves of the main feedwater systems in a PWR, which can lead to increased local flow velocity. This has the potential for causing increased FIC. Throttling with corresponding differential pressure can lead to the occurrence of two-phase conditions with a redistribution of the alkalizing agent, and, ultimately, increased iron solubility can be observed. Figure III–3 shows the corresponding results at various sensitive locations in a German PWR, with the increased relative wear rate depending on the operating partial power level. The scale of FIC wall thinning strongly depends on the potential for chemical dissolution of the oxide film. The pH value has a significant impact on the solubility of magnetite, because high pH values significantly reduce the potential for chemical dissolution of iron. It is generally accepted that a higher pH will reduce the FIC rate and that a pH at 25°C > 9.7 reduces FIC significantly. In two-phase flow, the relevant pH value is the pH of the water phase, considering the volatility coefficient between the water and steam phase for the respective alkalizing agent.

Such an effect is strongly dependent on the individual load case; therefore, it is necessary to evaluate the degradation behaviour of components and equipment and its predictability for each particular plant and for various planned power levels.

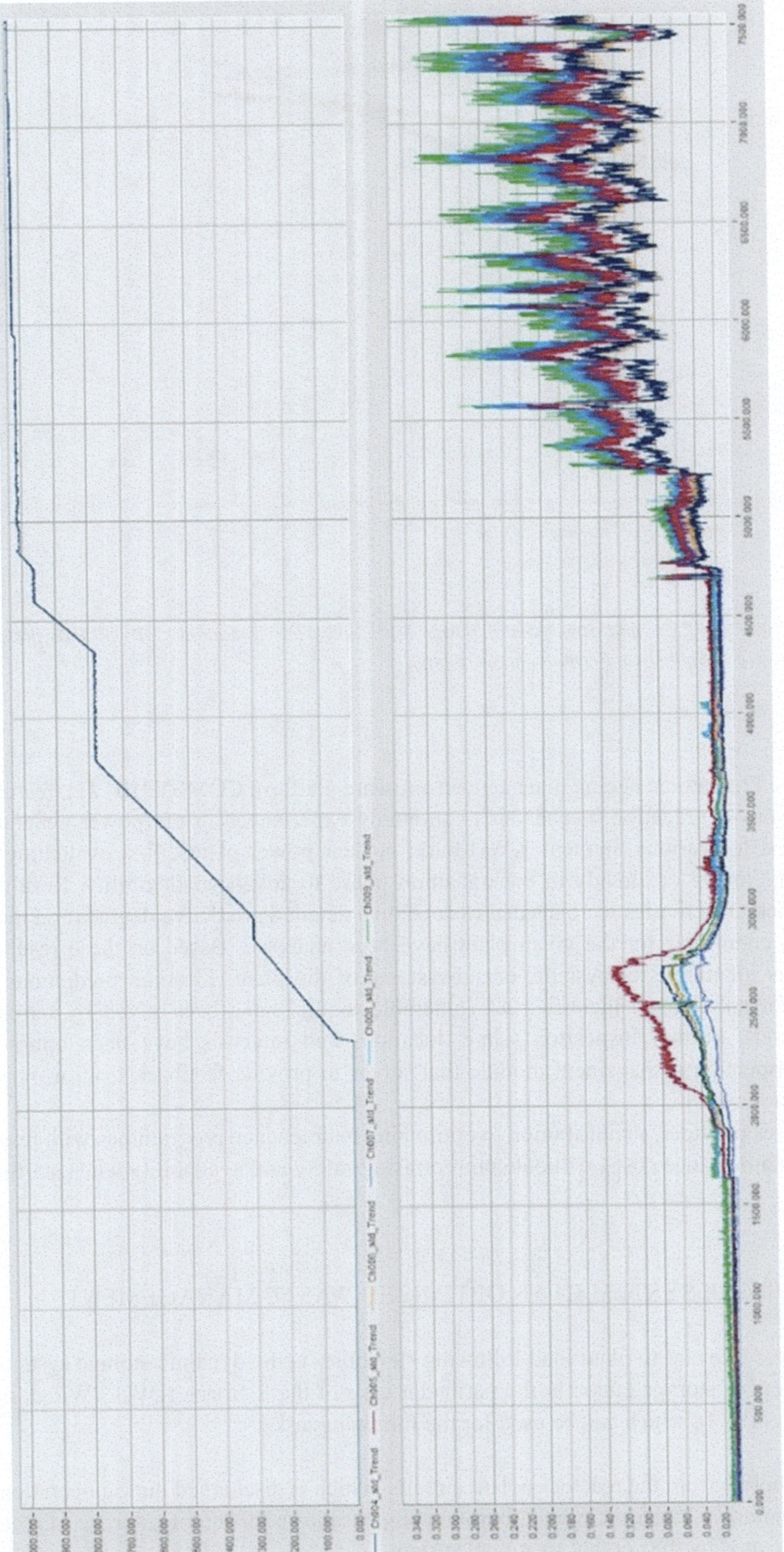

FIG. III–2. Non-stationary vibration levels on condensate pipe during power transient (reproduced from Ref. [III–18] with permission).

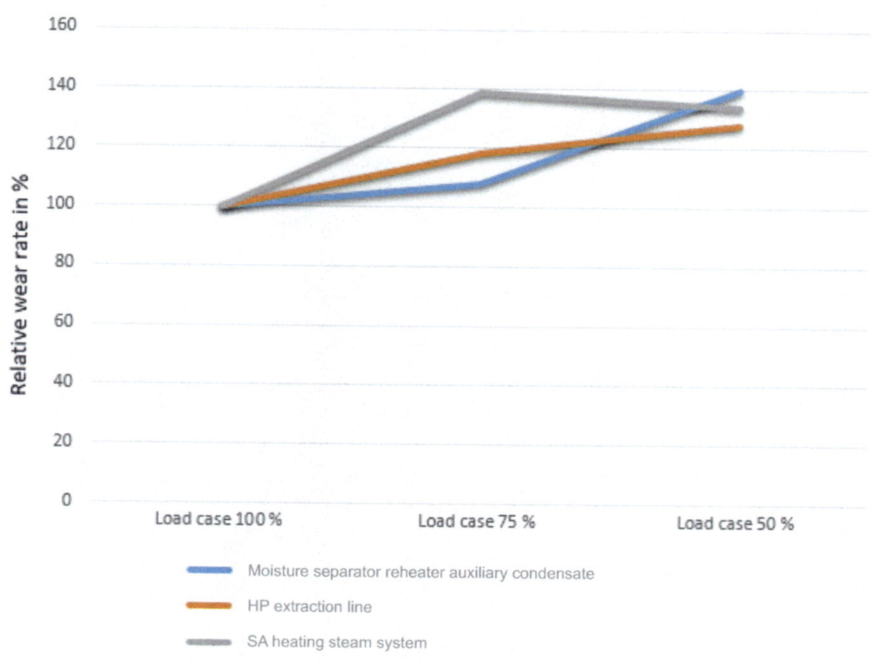

FIG. III–3. Relative wear rate for 75% and 50% partial power level at various locations. HP — high pressure; SA — steam auxiliary (courtesy of T. Salnikova, Framatome, Germany).

For this purpose, the Framatome ageing management software platform COMSY [III–21], featuring lifetime analysis for various degradation mechanisms commonly experienced in the power generation industry, has been applied for various Siemens KWU-build nuclear power plants, first evaluating the overall business case for FlexOp or already in the transition phase to advanced flexibility. Results of more than 30 years of research activities were needed to develop a detailed predictive degradation model for FIC. Water chemical conditions for the given plant have been analysed. Based on these results, a degradation potential was identified for systems or subsystems of the plant. Lifetime predictions for individual components for subsystems identified as vulnerable to FIC have been given by terms of performed detailed analysis. Further inspection scope, locations and intervals have been optimized applying an integrated inspection management module that serves to provide feedback from in-service inspection (ISI) examination results.

Overall, such analyses provided a contribution to optimizing maintenance programmes with respect to FlexOp and reducing maintenance costs, without compromising safety and availability when starting a new mode of operation.

III–3. DYNAMIC STORAGE SYSTEM IN AN OPTIMIZED WASTE MANAGEMENT

One additional feature relevant to plant load following flexibility is the dynamic storage system. It consists of various numbers of storage tanks; in the particular case of the Siemens KWU PWR design, there are six tanks (see Fig. III–4), which can be used for the following tasks:

— Receipt of reactor coolant from the reactor coolant circuits, which is discharged during operation of the plant (startup, shutdown, changes in load, burnup compensation), for interim storage of reactor coolant before it is treated by the coolant treatment system;

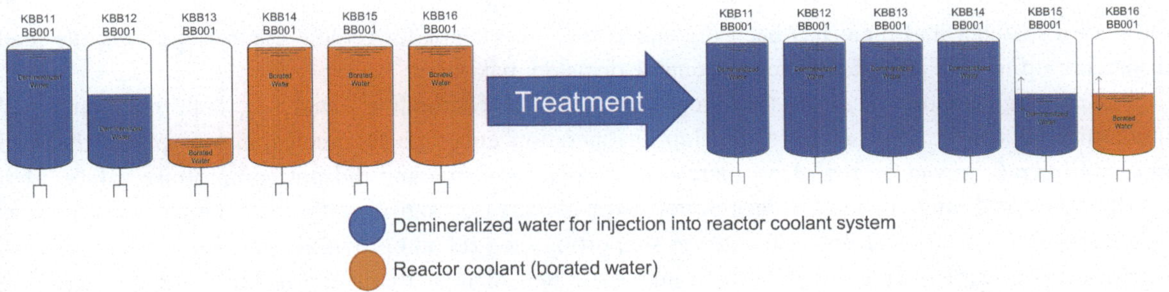

FIG. III–4. Scheme of the dynamic storage system (courtesy of T. Salnikova, Framatome, Germany).

— Storage and supply of demineralized water in sufficient quantity for operational needs (e.g. for dilution of reactor coolant).

The automatic control of the dynamic coolant storage contains the following two separate automatic step programmes (subgroup controls) to control filling and draining of the tanks in a predetermined order:

— Subgroup control for storage of borated water (coolant);
— Subgroup control for storage of demineralized water.

In such a design for dynamic coolant storage, borated water is filled subsequently into tanks 6, 5, 4, etc. and is drained from the tanks (mainly to the coolant treatment system) in reverse order. Demineralized water is injected subsequently into tanks 1, 2, 3, etc. and is drained from the tanks (mainly to the reactor boron and water make-up system) in reverse order.

During normal operation, both subgroup controls are in operation to enable simultaneous storage (i.e. receipt and release) of both borated water and demineralized water without any mixing. One tank is connected to the borated water line (mainly tank 6 is preselected) while another tank is connected to the demineralized water line (mainly tank 1 is preselected), thus giving the opportunity of storing and delivering the medium at the same time. The medium preselection of a tank (i.e. demineralized water or borated water) is performed by a selector according to the position of the valves connecting the demineralized water line and borated water line.

The dynamic coolant storage system is installed in all Siemens KWU-build PWR plants and has been adopted in EPR reactors as well. Retrofitting a dynamic storage system could be feasible for other existing nuclear power plants, but the availability of space and the cost need to be evaluated in detail. For new build designs, such a solution is considered to be an attractive option because the total necessary volume for storage of coolant and demineralized water can be reduced by the dynamic storage system.

Overall, such systems allow operation also with high demineralized water demand during the load following operation and avoid limitations related to water storage.

III–4. OPERATIONAL EXPERIENCE AND CONCLUSIONS

In 2021, transition aspects of the Siemens KWU-build nuclear power plants from limited FlexOp to advanced flexible performance in fully automatic mode were analysed in detail [III–22]. In addition to original 'build-in' features, the importance of further development and application of I&C and other digital solutions was demonstrated. Overall optimizations, including power manoeuvring guidelines, fuel management strategies, plant operational procedures, appropriate monitoring, and maintenance concepts, showed advantages for normal operation and allowed nuclear power plants to move towards advanced flexibility. Precise core monitoring, simulation and prediction tools provide the support needed to perform power ramps for both PWRs and BWRs. Furthermore, clear and transparent communication between the load dispatcher and plant has been organized. The key factor for remote control is giving the load dispatcher up-to-date information for reasonable decision making.

Applying lessons learned and carrying out corresponding analyses, optimizations and modernizations made it possible to develop an operation envelope for nuclear power plants that had minimal overall impact on the plant, increased the plant revenue and did not compromise safety. As a result, the overall range of the services performed has increased significantly over the past two decades (see Ref. [III–23]). Nuclear power plants are supporting the grid and providing various services, such as frequency containment reserve (FCR) and automatic and manual frequency restoration reserve (aFRR and mFRR). Economic dispatch (as a result of portfolio optimization) and overload management are also an everyday business for the nuclear power plants, together with emergency load reductions and voltage regulation, providing reactive power. A survey on performed grid services by German nuclear power plants can be found in table 13 of Ref. [III–24]).

For PWRs, FlexOp grid services have been mostly performed in the power range between 100% and 50–60% REO (in the power range of the constant average cooling temperature). For BWRs, FlexOp was mostly provided by the speed controlled internal recirculation pumps with a typical minimal load of 70% REO (without rod movements).

The introduction of digital I&C and optimizations of core control with ALFC (Section III–1.3) were reported by German PWRs Brokdorf, Grohnde and Isar 2 to be the most beneficial of the modernizations. The Isar 2 plant serves as an example, providing various grid services within a band of approximately 590 MW(e) in fully automatic remote controlled mode. Improved technology has allowed the plant to reach both targets of increasing the range of services performed (see Ref. [III–24]) and improving reliability over the years.

FlexOp modes were also optimized among other advantages with respect to smooth plant performance, including minimizing the injection of borated or demineralized water and thus reducing activation of the CVCS. The ramp rate to achieve was up to 40 MW(e)/min, and the modernizations should allow FlexOp without manual intervention in the whole range between minimum load and 100%.

Looking at the number of reported events in this nuclear power plant during the decade with advanced FlexOp, a decrease can be seen in Ref. [III–24], providing the best proof of the chosen overall plant concept for FlexOp. Most of the solutions can be adapted to other plant designs.

REFERENCES TO ANNEX III

[III–1]   LUDWIG, H., SALNIKOVA, T., STOCKMAN, A., WAAS, U., Load cycling capabilities of German nuclear power plants (NPP), ATW-Int. J. Nucl. Power **55** 8/9 (2010).

[III–2]   INTERNATIONAL ATOMIC ENERGY AGENCY, Non-baseload Operation in Nuclear Power Plants: Load Following and Frequency Control Modes of Flexible Operation, IAEA Nuclear Energy Series No. NP-T-3.23, IAEA, Vienna (2018).

[III–3]   INTERNATIONAL ATOMIC ENERGY AGENCY, Progress on Pellet–Cladding Interaction and Stress Corrosion Cracking, IAEA-TECDOC-1960, IAEA, Vienna (2021).

[III–4]   KUHN, A., KLAUS, P., Improving automated load flexibility of nuclear plants with ALFC, VGB PowerTech **96** 5 (2016) 48–52.

[III–5]   KUHN, A., SCHIRRMEISTER, K., Advancing load following control, Nucl. Eng. Int. (Jan. 2018), https://www.neimagazine.com/advanced-reactorsfusion/advancing-load-following-control-6015633/

[III–6]   MOROKHOVSKYI, V., Reactor core control based on artificial intelligence, ATW-Int. J. Nucl. Power **65** 6/7 (2020) 350–352.

[III–7]   SALNIKOVA, T., MOROKHOVSKYI, V., RUDOLPH, J., "Transfer of concepts for flexible operation to different plant types", paper presented at Annual Mtg on Nuclear Technology, Berlin, 2017.

[III–8]   MOROKHOVSKYI, V., "Reactor core control based on artificial intelligence", paper presented at Int. Conf. on Nuclear Engineering & ASME Power Conf. (ICONE-POWER 2020), Anaheim, CA, 2020.

[III–9]   RUDOLPH, J., BERGHOLZ, S., HEINZ, B., JOUAN, B., "AREVA fatigue concept – a three stage approach to the fatigue assessment of power plant components", Nuclear Power Plants (CHANG, S.H., Ed.), IntechOpen, London (2012) Ch. 11,
https://cdn.intechopen.com/pdfs/33375/InTech-Areva_fatigue_concept_a_three_stage_approach_to_the_fatigue_assessment_of_power_plant_components.pdf

[III–10] JOUAN, B., BERGHOLZ, S., RUDOLPH, J., KÖNIG, G., MANKE, A., "Automatic fatigue monitoring based on real loads -calculation example of a flange", Proc. ASME 2013, Paris, 2013, ASME, New York (2013) PVP2013-97660,
https://dx.doi.org/10.1115/PVP2013-97660

[III–11] JOUAN, B., RUDOLPH, J., BERGHOLZ, S., "Fatigue monitoring system and post-processing of temperature measurements: Surge line under stratification loading", Proc. ASME 2015 Pressure Vessels and Piping Conf., Boston, MA, 2015, ASME, New York (2015) PVP2015-45676
https://dx.doi.org/10.1115/PVP2015-45676

[III–12] RUDOLPH, J., BERGHOLZ, S., VORMWALD, M., BAUERBACH, K., On the methodology of fatigue assessment of nuclear power plant components, Mater. Test. **53** 7/8 (2011) 407–417.

[III–13] MOUSSALAM, N., ZIEGLER, R., RUDOLPH, J., BERGHOLZ, S., A method for monitoring vibrational fatigue of structures and components, J. Pressure Vessel Technol. **144** 3 (2022) 031304,
https://doi.org/10.1115/1.4053380

[III–14] LE DUFF, J.A., et al., "High cycle thermal fatigue issues in RHRS mixing tees and thermal fatigue test on a representative 304 L mixing zone", Proc. ASME 2011 Pressure Vessels and Piping Conf., Baltimore, MD, 2011, ASME, New York (2011) PVP2011-57951, 691–699.

[III–15] IMBROGNO, G., MARLETTE, S., CAROLAN, A., UDYAWAR, A., GRAY, M., "Recent operational experience of pressurized water reactor safety injection and drain line cracking and supporting flaw evaluations", Proc. ASME 2019 Pressure Vessels and Piping Conf., San Antonio, TX, 2019, ASME, New York (2019) PVP2019-93945.

[III–16] TREWIN, R., RUDOLPH, J., "Fast-running thermal-hydraulic computer program for screening branch pipes subject to thermal fatigue", paper presented at 19th Int. Topical Mtg on Nuclear Reactor Thermal Hydraulics (NURETH-19), Brussels, 2022.

[III–17] FOMI WAMBA, F., et al., "Operation optimization using data analytics CADIS -a Framatome solution", paper presented at 39th Annual Conf. of the Canadian Nuclear Society and 43rd Annual CNS/CNA Student Conf., Ottawa, 2019.

[III–18] SALNIKOVA, T., et al., Survey on Vibration Impact of Flexible Operation on NPP: Based on Operational Experience, Energiforsk Report 2021:774, Energiforsk, Stockholm (2021),
https://energiforsk.se/media/29825/survey-on-vibration-impact-of-flexible-operation-on-npp-energiforskrapport-2021-774.pdf

[III–19] HEITMANN, G., KASTNER, W., Erosionskorrosion in Wasser-Dampfkreisläufen -Ursachen und Gegenmaßnahmen, VGB Kraftwerkstech. **62** (1982).

[III–20] ZANDER, A., NOPPER, H., Evaluation of wall thinning affects caused by superposition of flow-induced degradation effects with the COMSY code, PPCHEM **12** 8 (2010) 494–499.

[III–21] ZANDER, A., NOPPER, H., ROESSNER, R., COMSY – Das Softwareprodukt für ein effizientes Lebensdauer- und Inspektionsmanagement, VGB PowerTech **85** 9 (2005) 126–131.

[III–22] SALNIKOVA, T., et al., Survey on Operational Experiences of NPPs in a Transitioning Energy System Going from Baseload to Flexible Operation – Germany Case Study, Energiforsk Report 2022, Energiforsk, Stockholm (2022).

[III–23] MÜLLER, C., "Operation experience in flexible mode, NPP ISAR 2, Preussen Elektra", Presentation at the TÜV Forum Kerntechnik, Berlin, 2022.

[III–24] HÄNNINEN, S., et al., Survey on Power System Ancillary Services, Energiforsk Report 2020:708, Energiforsk, Stockholm (2020),
https://energiforsk.se/media/28856/survey-on-power-system-ancillary-services-energiforskrapport-2020-708.pdf

ON-LINE MONITORING SUCCESS STORY: EXTENDING THE CALIBRATION INTERVALS OF NUCLEAR PLANT PRESSURE TRANSMITTERS

On-line monitoring (OLM) to verify the calibration of nuclear plant pressure, level and flow transmitters received regulatory approval in 2005 from the British authorities and has been used at the Sizewell B plant in the United Kingdom for nearly two decades with resounding success. The technology has saved Sizewell millions of pounds sterling per operating cycle in personnel time, outage time and radiation exposure. The method was approved by the US authorities in 2021 and is expected to receive widespread implementation almost immediately. In fact, the Southern Nuclear Operating Company is working with Analysis and Measurement Services Corporation (AMS) to obtain a License Amendment for the US Nuclear Regulatory Commission (NRC) to implement OLM in its fleet of eight nuclear power units. The OLM methodology has also been used in the French fleet and is poised for implementation in nuclear plants in China, the Republic of Korea, the Russian Federation, Ukraine and elsewhere.

OLM technology for monitoring transmitter drift is based on a comparison of each redundant transmitter's reading with the average reading of its redundant group. The average reading, which is referred to as 'process estimate', is calculated using 'simple' and 'parity space' averaging techniques or empirical and physical modelling. Plenty of references are publicly available on how OLM data analysis may be performed to verify the calibration of pressure transmitters.

Some of the information provided in this annex has been previously published by H.M. Hashemian in the American Nuclear Society's *Nuclear News* magazine [IV–1].

IV–1. CONVENTIONAL CALIBRATION VERSUS ON-LINE MONITORING

Conventional calibrations of transmitters typically involve two steps:

— Step 1: Determine whether the transmitter needs to be calibrated. This step is carried out by manually isolating the transmitter from the process and applying a range of known pressures to the transmitter covering the operating range of the transmitter, while measuring its output. The data from this step are referred to as the 'as-found' calibration data. If the 'as-found' data show that the calibration of a transmitter is acceptable, then no further action is needed, and the transmitter is returned to service. Otherwise, the transmitter is calibrated as detailed in the next step.
— Step 2: Calibrate the transmitter. This step is carried out by making manual adjustments to transmitter zero and/or span settings via on-board potentiometers to make the transmitter read a range of applied pressures as closely as possible. The data from this step are referred to as 'as-left' calibration data.

The first step above consumes the majority of the effort that is expended by plant personnel on sensor calibrations that can be saved by OLM. In particular, a review of calibration history of nuclear plant pressure, level and flow transmitters has shown that about 90% of these transmitters maintain their calibration for much longer than a typical fuel cycle that can range from 14 to 24 months. More specifically, calibration records have shown that only about 10% of nuclear plant transmitters exceed their 'as-found' limits [IV–2]. Accordingly, the OLM procedure was developed to eliminate about 90% of the calibration burden on nuclear plant personnel, while improving plant safety and efficiency by eliminating unnecessary calibrations that can cause damage to plant equipment, expose plant personnel to radiation and increase the potential for human error, plant trips or spurious actuations.

IV–2. ON-LINE MONITORING BASICS FOR TRANSMITTER CALIBRATION MONITORING[1]

OLM technology for transmitter drift monitoring involves a simple procedure that is passive and benign to plant operation and does not require any modification to the plant. All that is needed to implement OLM is a means to retrieve the output readings of transmitters, which can be accomplished using the plant computer or a separate data acquisition system and a software package to validate and analyse the transmitter data. OLM is not a substitute for conventional calibrations. Rather, it is an analytical tool analogous to using measuring and test equipment (M&TE) to check for drift of transmitters during plant operation in order to determine whether they have to be scheduled for a physical calibration by plant personnel during an upcoming plant outage.

To perform OLM, readings of redundant sensors are tracked while the plant is operating to identify drift beyond acceptable limits. Figure IV–1 shows the readings of four redundant steam generator level transmitters at Unit 2 of the McGuire Nuclear Station over a period of about 30 months, representing nearly two full operating cycles. A multichannel data acquisition system was installed at McGuire and connected to existing plant isolation modules to allow live data collection while the plant was operating. The work was done over the period of 1992 to 1995 in collaboration with Duke Power Company, the owner of the McGuire Nuclear Station. The results are documented in NUREG/CR-5903 (1993) [IV–3] and NUREG/CR-6343 (1995) [IV–4].

To obtain the deviation plot in Fig. IV–1, an estimate of the true steam generator level was first obtained by averaging the four signals. Next, the process estimate was subtracted from the reading of each transmitter to yield the deviation of each transmitter from the average. The OLM limits for the steam

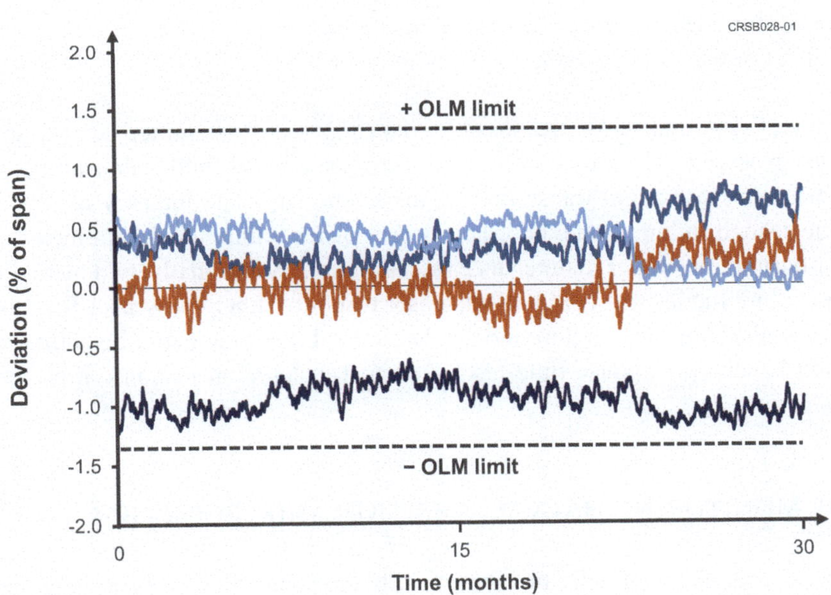

FIG. IV–1. OLM data for four redundant steam generator level transmitters. (Source of data: McGuire Nuclear Station Unit 2 — NUREG/CR-6343 [IV–4].)

---

[1] Information from this section was previously published by H.M. Hashemian in the American Nuclear Society's *Nuclear News* magazine [IV–1].

generator level transmitters, shown as dotted lines in Fig. IV–1, were calculated based on plant set point methodology and set point data [IV–4].

It is obvious from the four traces in Fig. IV–1 that the four McGuire transmitters did not drift over the two operating cycles shown in the figure and in fact remained well within the plant's OLM limits. With this information, it is reasonable to claim that the calibrations of these transmitters are intact. However, it is important to note that this claim would be true only if the four transmitters did not all drift together in either the positive or negative direction (i.e. common mode drift). If there is no common mode drift, then their average value is a close representation of the true process. The potential for common mode drift has been researched extensively and ruled out for the existing generation of nuclear grade pressure, level and flow transmitters.

IV–3. DETECTING TRANSMITTER FAILURE MODES WITH ON-LINE MONITORING

Two major failure mode and effects analysis (FMEA) processes have been performed by the Electric Power Research Institute (EPRI) on calibration and response time of nuclear grade pressure, level and flow transmitters in nuclear power plants [IV–5, IV–6]. As indicated in these EPRI studies, OLM can reveal performance failure modes of pressure, level and flow transmitters, except for a few that affect response time.

The EPRI report 'Investigation of Response Time Testing Requirements' [IV–5] examined 14 pressure transmitters from five manufacturers to determine what failure modes were detectable by response time testing and likely to occur during service in a nuclear power plant. In all, FMEA identified two failure modes that would affect a transmitter response time while not concurrently impacting its calibration:

— Slow sensor fill fluid leaks during pressurized operation;
— Poor adjustment of the variable damping potentiometer.

Various types of fill fluid leaks are possible, depending on transmitter design, seal failures and pressure conditions in service. The only confirmed occurrence of a fill fluid leak leading to response time degradation has been in Rosemount transmitters. This is known in the industry as the 'Rosemount oil loss' problem. The second failure mode, poor adjustment of the damping potentiometer, can only occur from human error during transmitter maintenance or calibration. Additionally, two manufacturing defects were identified as failure modes that could affect sensor response time: low sensor fill fluid and crimped capillary lines. An analysis of these failure modes determined that they could be addressed using either post-manufacturing benchtop response time testing or post-installation response time testing using the noise analysis technique.

IV–4. ON-LINE MONITORING DATA ACQUISITION AND PROCESSING

OLM relies on a sequence of activities that include the acquisition of plant data, the qualification of those data to ensure suitability for use, and the analysis needed to establish reliable estimates of sensor performance.

IV–4.1. On-line monitoring data acquisition

Nuclear power plants are often equipped with the means to continuously collect and store the outputs of their process sensors, which can be retrieved from the plant computer or through its data historian. Typically, the sensor output readings are converted to engineering units, time stamped and stored as shown in Fig. IV–2 [IV–7] for three redundant level transmitters in a pressurized water reactor (PWR)

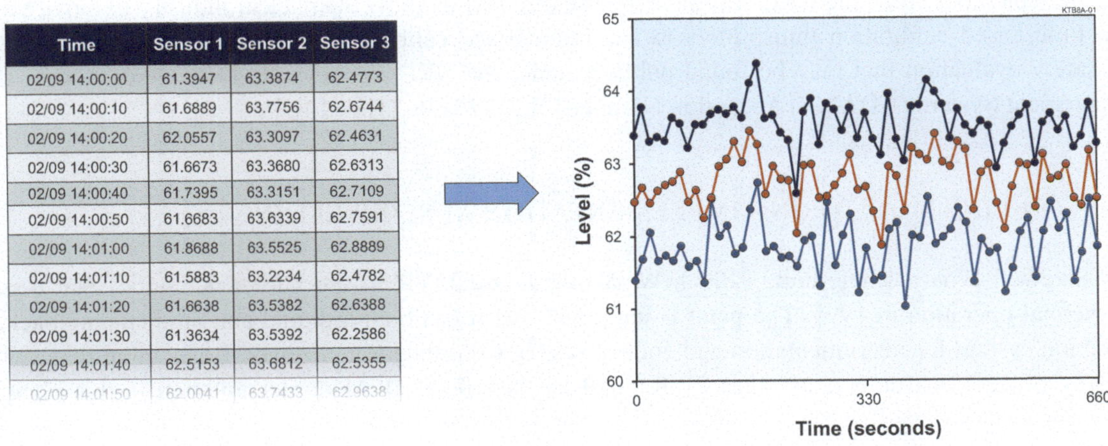

FIG. IV–2. OLM data from a PWR plant computer (reproduced from Ref. [IV–7]).

plant. If the data are not available from the plant computer or its historian, a separate data acquisition system has to be used to acquire the OLM data. The sampling rate and duration of OLM data acquisition are important to the reliability of its results and need to be selected carefully to capture all the information that is needed to establish the performance of the sensors.

IV–4.2. On-line monitoring data qualification

The raw data from the plant computer or its data historian are not usually ready for analysis immediately after they are retrieved. The data can be compressed or have areas of missing regions or stuck data values, spikes, noise or other anomalies. These effects are normally benign to the plant operation and safety but have to be addressed to qualify the OLM data for analysis.

IV–4.3. On-line monitoring data analysis

Analysis of OLM data requires a reference value as the basis for detecting drift. The reference value is also referred to as the process estimate and can be calculated using a variety of averaging and modelling techniques.

IV–5. ON-LINE MONITORING IMPLEMENTATION IN US NUCLEAR POWER PLANTS

Over the past 15 years, OLM has been implemented in the following US nuclear power plants on an experimental basis:

— Watts Bar Nuclear Plant Unit 1 (a 4-loop Westinghouse PWR): Transmitters were monitored for one cycle from November 2006 to February 2008 [IV–8].
— Joseph M. Farley Nuclear Plant Units 1 and 2 (3-loop Westinghouse PWRs): Transmitters were monitored over multiple cycles from April 2008 to July 2011 [IV–9].
— North Anna Power Station Units 1 and 2 (3-loop Westinghouse PWRs): Transmitters were monitored over multiple cycles from January 2008 to April 2011 [IV–9].

In 2018, the Alvin W. Vogtle Electric Generating Plant Units 1 and 2 (4-loop Westinghouse PWRs) began using OLM on a non-experimental basis; transmitters have been monitored from October 2018 to the present as part of an ongoing commercial OLM implementation [IV–10] to [IV–12].

In September 2021, the NRC approved the use of OLM to allow the nuclear industry to move away from time based calibration transmitters to condition based calibrations. The NRC approval is given in a safety evaluation that may be found publicly under the NRC Agencywide Documents Access and Management System (ADAMS) Accession Number ML21179A062 [IV–13].

IV–6. ON-LINE MONITORING IMPLEMENTATION AT SIZEWELL B[2]

Sizewell B is a single-unit, 1200 MW Westinghouse PWR in the United Kingdom that began commercial operation in 1995. The plant is unique in that it has both a digital and an analogue backup protection system for instrumentation and control (I&C). Consequently, Sizewell B has more than twice as many process instruments as other PWRs, and the benefit of OLM implementation in this plant is very significant.

OLM implementation at Sizewell B over the period of 2005 to 2020 involved 197 transmitters, producing a huge database of OLM results. For example, there are 435 cases in the database involving 108 transmitters that were monitored over five operating cycles by OLM and subsequently calibrated, providing the opportunity to compare the OLM results with manual calibrations. A summary of this comparison is provided in Table IV–1.

The key conclusions are as follows:

— The OLM and manual calibration results for 356 of 435 transmitters (over 80%) matched perfectly. Although OLM and manual calibrations are not exactly the same, due to the effect of process conditions and the different number of components that are involved in the two tests, this good agreement is nevertheless important, as it provides confidence in the validity of OLM technology.
— For 77 transmitters (nearly 18%), OLM found the transmitters as having drifted beyond their OLM limits while manual calibrations showed no significant drift. Although the two methods did not produce comparable results, this outcome is readily acceptable because it is conservative.
— OLM did not flag two transmitters that were found to be bad by manual calibrations. The cause of this discrepancy could not be determined. Upon arrival at this outcome, which is not conservative, the Sizewell B engineers compared this observation with their experience with discrepancies in manual calibrations over the years since 1996. This effort showed that Sizewell B has experienced an average of three discrepancies due to human error and miscalibrations per each operating cycle [IV–14]. Therefore, Sizewell B engineers concluded that the two discrepancies seen here are readily acceptable because they are better than the conventional practice, where about 15 cases of human error and miscalibrations would have typically occurred over the same period. With only two non-conservative results out of 435 cases, it is reasonable to conclude that OLM correctly or conservatively identified greater than 99% of the Sizewell B transmitters that needed a calibration check.
— In a 2019 update, it became known that Sizewell is in possession of 921 cases of which 11 are non-conservative. Again, this statistic confirms that OLM has correctly or conservatively identified about 99% of the transmitters that needed a calibration check [IV–1].

The EPRI study [IV–15] supports the conclusion that OLM fails to identify drifting transmitters in less than 1% of cases, which is better than the approximately 5% failure rate associated with traditional calibrations.

---

[2] Information from this section was previously published by H.M. Hashemian in the American Nuclear Society's *Nuclear News* magazine [IV–1].

TABLE IV–1. AGREEMENT BETWEEN OLM AND CALIBRATIONS FOR SIZEWELL B TRANSMITTERS

OLM	Calibration	Number of matches	Assessment
Good	Good	332	Perfect match
Bad	Bad	24	Perfect match
Bad	Good	77	Conservative mismatch
Good	Bad	2	Non-conservative mismatch

REFERENCES TO ANNEX IV

[IV–1] HASHEMIAN, H.M., Online monitoring technology to extend calibration intervals of nuclear plant pressure transmitters, Nuclear News **64** 7 (2021).

[IV–2] WOOTEN, B., Instrument Calibration and Monitoring Program, Vol. 1: Basis for the Method, TR-103436-V1, Electric Power Research Institute, Palo Alto, CA (1993).

[IV–3] HASHEMIAN, H.M., et al., Validation of Smart Sensor Technologies for Instrument Calibration Reduction in Power Plants, NUREG/CR-5903, Nuclear Regulatory Commission, Washington, DC (1993).

[IV–4] HASHEMIAN, H.M., On-Line Testing of Calibration of Process Instrumentation Channels in Nuclear Power Plants, NUREG/CR-6343, Nuclear Regulatory Commission, Washington, DC (1995).

[IV–5] SWISHER, V., Investigation of Response Time Testing Requirements, NP-7243-R1, Electric Power Research Institute, Palo Alto, CA (1994).

[IV–6] WOOTEN, B., Instrument Calibration and Monitoring Program, Vol. 2: Failure Modes and Effects Analysis, TR-103436-V2, Electric Power Research Institute, Palo Alto, CA (1993).

[IV–7] HASHEMIAN, H.M., SHUMAKER, B.D., MORTON, G.W., Online Monitoring Technology to Extend Calibration Intervals of Nuclear Plant Pressure Transmitters, AMS Topical Report AMS-TR-0720R2-A, NRC ADAMS Accession No. ML21235A493, Analysis and Measurement Services Corporation, Knoxville, TN (2021).

[IV–8] HASHEMIAN, H.M., On-Line Calibration Monitoring of Safety-Related Pressure Transmitters at Watts Bar Unit 1, EPRI Final Report, Electric Power Research Institute, Palo Alto, CA (2010).

[IV–9] HASHEMIAN, H.M., SHUMAKER, B., MORTON, G., CAYLOR, S., On-line Monitoring of Accuracy and Reliability of Instrumentation and Health of Nuclear Power Plants, Phase II+ Final Report, Vols 1 and 2, Report No. DOE/ER84626, DOE Grant No. DE-FG02-06ER84626, USDOE, Washington, DC (2011).

[IV–10] MORTON, G., et al., Results of Mid-cycle Analysis of On-line Calibration Monitoring Data for Pressure Transmitters at Vogtle Unit 1 from October 2018 through June 2019, AMS Report VOG1905R0, AMS Corporation, Knoxville, TN (2019).

[IV–11] SHUMAKER, B., et al., Results of Mid-cycle Analysis of On-line Calibration Monitoring Data for Pressure Transmitters at Vogtle Unit 2 from March 2019 through November 2019, AMS Report VOG1906R0, AMS Corporation, Knoxville, TN (2019).

[IV–12] MORTON, G., et al., Results of Full Cycle Analysis of On-line Calibration Monitoring Data for Pressure Transmitters at Vogtle Unit 1 from October 2018 through March 2020, AMS Report VOG2005R0, AMS Corporation, Knoxville, TN (2020).

[IV–13] NUCLEAR REGULATORY COMMISSION, Safety Evaluation by the U.S. Nuclear Regulatory Commission for Analysis and Measurement Services Corporation Topical Report AMS-TR-0720R1, "Online Monitoring Technology to Extend Calibration Intervals of Nuclear Plant Pressure Transmitters" by the Office of Nuclear Reactor Regulation (EPID No. L-2020-TOP-0037), NRC ADAMS Accession No. ML21179A062, NRC, Washington, DC (2021),
https://www.nrc.gov/docs/ML2117/ML21179A062.pdf

[IV–14] GOFFIN, P., Sensor Calibration Extension (EC109087) Additional Work to Support Continued Implementation, Sizewell B Power Station, Systems Engineering, SZB/ESR/503 (2019).

[IV–15] HASHEMIAN, H.M., RASMUSSEN, B., Plant Application of On-Line Monitoring for Calibration Interval Extension of Safety-Related Instruments: Vols 1 and 2, EPRI-TR-1013486, Electric Power Research Institute, Palo Alto, CA (2006).

Annex V

**HOW DIGITAL TOOLS CAN SUPPORT THE NUCLEAR
INDUSTRY — TRACTEBEL DIGITAL TWIN CASE STUDIES**

V–1. INTRODUCTION

Tractebel's operational expertise has been accrued from four decades as the architect and responsible designer of the Belgian nuclear power plants, which are owned and operated by its parent company, Engie. This expertise helps Tractebel to anticipate the needs of the nuclear industry needs and drives its internal research, development and innovation to adapt its skills and tools to the evolving market. This strategy allows Tractebel to tackle complex and first of a kind challenges in the most pragmatic and cost effective way.

The objective of this annex is to illustrate, with a selection of practical examples from Tractebel projects, how digital twins' functionalities and their development approach may increase the efficiency and performance of the nuclear industry.

V–2. DIGITAL TWIN DEFINITION

There are currently various perceptions and definitions in the nuclear industry of a digital twin. The US Nuclear Regulatory Commission (NRC) defines a digital twin system as the overall system that comprises at least the physical system, the virtual system and the relationships between physical and virtual systems (e.g. information and data flows [V–1]). The NRC also adds three conditions for any existing or new technology to be referred to as, or to be a part of, a nuclear digital twin:

— Digital form: The technology should exist in a digital form that can be managed, processed and executed on a digital device.
— State concurrency: The technology should be capable of updating dynamically and in real time to represent the state of a physical entity or physical phenomenon and should maintain that state concurrency.
— Purpose: The technology should have an underlying purpose related to a nuclear power plant life cycle activity. That is, a nuclear digital twin should be part of a digital twin system and be coupled in terms of both state and purpose with the physical nuclear power plant.

Another practical approach for defining a digital twin is to focus on the functionalities offered and their similarities and differences with building information modelling (BIM). Both digital twins and BIM are based on a common digital model. This digital model is a virtual representation of the physical facility. It should allow full collaboration on an integrated model accessible from a common data environment. The model comprises not only 2-D or 3-D geometries, but also the data of the facilities (its properties and quantities). The digital model is used to perform calculations and simulations and is linked to a document management system.

As stressed in the NRC definition of a digital twin, the emphasis is put on its relational interaction with the physical world (with sensors and monitoring systems), enabling the collection of dynamic/real time data. Performing analytics and advanced simulations on this flow of dynamic data (including artificial intelligence and machine learning) then makes it possible to generate predictions and recommendations for the operation of the physical asset.

Other themes are closely linked to BIM and digital twins and are often integrated with them: augmented reality (AR)/virtual reality (VR), systems engineering and specific tools for brownfield (existing asset) applications.

Compared with other technical fields and industries, the nuclear industry has some specific characteristics:

— The need for complex simulations: Simulation is integral to nuclear engineering, but developing a full scope digital twin at nuclear power plant scale necessitates the combination of small scale or component digital twins into interconnected systems digital twins.
— The extended use of data (static and dynamic) from heterogeneous systems with different fidelities and resolutions.
— The diversity of digital twin use cases and support functions: Engineering, safety and licensing, operation and maintenance training, diagnostics for predictive maintenance and optimal operation, and decision support for incident/accident management.
— Challenging centralization of legacy data from operating nuclear power plants (decentralized and/or unstructured databases).
— Limitations arising from confidentiality and accessibility rules.

Development efforts are currently needed to bridge the gaps in implementing a full scope digital twin as defined above.

In conclusion, it might take some time for a clear definition of a digital twin system and its technologies to emerge and gain consensus in the nuclear industry and among its technology providers. Whatever the exact definition of the concept, it is important to understand the key components and functions of any claimed or specific digital twin application.

V–3. DIGITAL TWIN FUNCTIONALITIES IN THE NUCLEAR INDUSTRY: TRACTEBEL CASE STUDIES

The following selection of practical examples from Tractebel's projects illustrates how digital twins' functionalities and their development approach may increase the efficiency and performance of the nuclear industry.

V–3.1. Digital models in simulation support of core monitoring and operation

Tractebel provides proficient engineering support to Electrabel, relying on its independent digital reactor modelling capacities. The core physics modelling backbone is the PANTHER code system, developed by EDF Energy and distributed by the ANSWERS Service (Winfrith, UK). This system is approved by the Belgian safety authorities for core design, in-core fuel management, reload safety demonstration, safety transient analysis for the functional safety analysis report and core operational support for the seven Belgian nuclear power plants. PANTHER is also a key component of the Tractebel CERES multiscale multiphysics platform, to be coupled for increased fidelity of complex physical phenomena modelling with codes of the COBRA family for sub-channel thermohydraulics or TRACE/RELAP for system thermohydraulics.

Beyond the use of the digital models of the Belgian reactors in order to provide studies for margin management, along with operational flexibility and functional safety analysis report assessment of evolving safety requirements, the same tools also drive tailored applications for direct core operational support of the utility. These applications fall into two categories: reactor connected applications, using digital models based on real time plant data (close to a digital twin of the core), and stand alone (off-line) applications, used for in-site management of the core power distribution follow-up and verification of safety parameters.

Tractebel's CORMORAN application for core monitoring and reactor analysis belongs to the first category and is installed at two sites (Doel and Tihange), coupled to the plant data acquisition system of each unit. It integrates two distinct functionalities in a single application using PANTHER core models:

— Core monitoring, addressing real time core simulation based on measured reactor data, allowing realistic core status display and best estimate transient predictive capability for operation strategy forecasting;
— In-core flux map processing, related to nuclear safety as it addresses the monthly regulatory power distribution analyses based on in-core flux measurements.

As part of the second category, the single point calibration method is used to calibrate ex-core power detectors from a single measured flux map ('single point'), completed by a simulated axial offset swing replacing the real plant swing manoeuvre with the associated necessary successive flux map measurements.

Tractebel digital reactor modelling capacities give Electrabel direct access to a wide range of independent engineering support activities to optimize the operation of its reactors both technically and economically. Beyond contributing to enhancing plant performance, the digital models of the Belgian reactors are also valuable assets because they ensure:

— Consistent quality of core physics simulation capabilities, validated for a wide range of engineering applications and based on 3-D digital models that are tailored in-house to the specific safety and operational needs of the utility;
— Cost and resource savings, based on available proven theoretical simulation expertise emphasized by sound historical knowledge of the plant licensing framework and its technical specificities;
— Enhanced quality assurance, theoretical knowledge and use proficiency, ensured as the neutronics engine is the same code used for reloading safety and accident analysis.

V–3.1.1. CORMORAN

CORMORAN takes the measured reactor data with a two minute sampling frequency as time dependent boundary conditions for the transient PANTHER core model providing the calculated evolution of the 3-D power distribution over successive time steps. All of the results are available on a local server for follow-up display, processing and (if necessary) operational forecast. The PANTHER core model is periodically self-calibrated to match the measured axial offset to ensure it duplicates as exactly as possible the real axial core power distribution.

The technical advantages offered by the enhanced core follow-up system and flux map processing are:

— Improvement of the capacity factor, with environmental and economic gains for the utility (e.g. limitation of liquid effluents due to dilution and boration with reduction of treatment costs, by optimizing power escalation or load manoeuvres for a fast return to full power).
— Capability to prevent operational penalties (e.g. temporary power limitation to cope for peaking factor limit overstepping) by providing a closer to reality core modelling of the in-core flux distribution taking into account previous history. This approach avoids strong measured core discrepancies from monthly flux when real core conditions differ from the static theoretical assumptions used for flux map analyses.
— Increase of core operational efficiency by early detection of possible operation anomalies and by optimization of corrective manoeuvres.

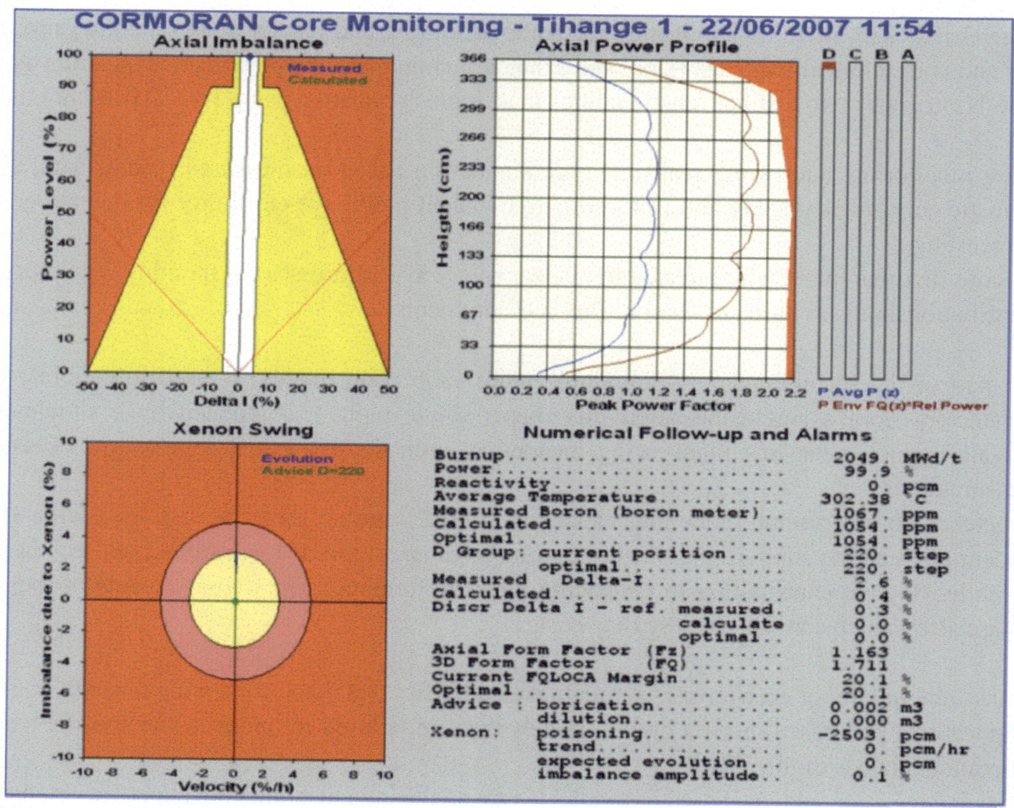

FIG. V–1. CORMORAN core status display (courtesy of C. Schneidesch, Tractebel Engineering, Belgium).

(a)   Core monitoring

The CORMORAN status is displayed in the operators' room (see Fig. V–1), periodically refreshed with the next data (update) acquisition and associated flux calculation. The CORMORAN display gives the operators the advantage of complete real time information on the core status, combining measured data and core parameters only accessible from simulation. The operators can:

— Easily monitor the current status of the core in terms of measured data: power level, average vessel temperature, boron concentration, the reactor axial offset indicator (ΔI), position of the ΔI in the authorized band, historical behaviour of the ΔI for the last six days.
— Access safety related core parameters that are not directly measured but are results of the theoretical calculation:
  • Axial power distribution and peak axial factor;
  • the peak fuel rod power factor (FQ) and remaining margin towards loss of coolant accident (LOCA);
  • Current xenon poisoning balance and expected evolution in the near future.
— Be informed of the configuration leading to an optimal situation.

(b)   Operation strategy forecast

Based on the best estimate core status, the second functionality of CORMORAN is used by the operators to predict, by simulation, the core behaviour in specific power manoeuvres, such as startup after shutdown, power transient, xenon power transient for the ex-core calibration, and return to full power after intermediate power state.

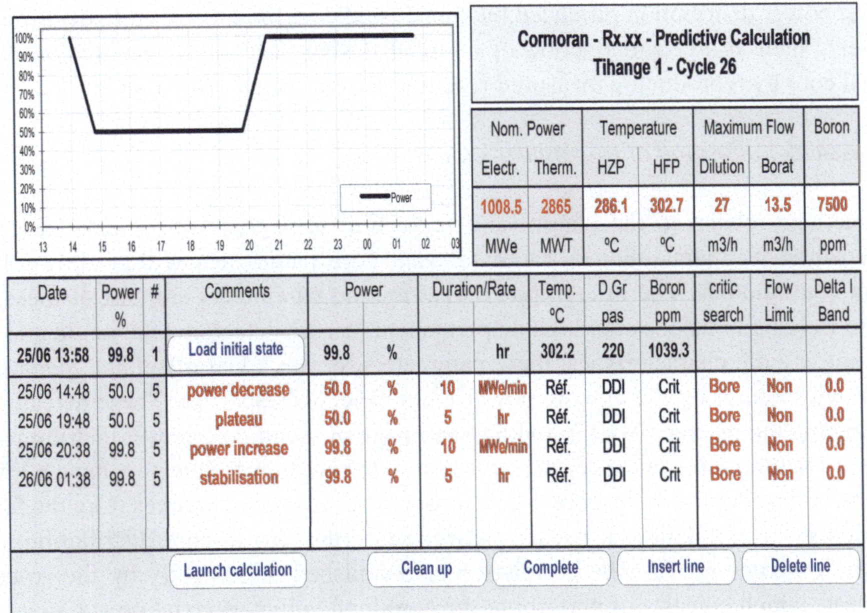

Nom. Power		Temperature		Maximum Flow		Boron
Electr.	Therm.	HZP	HFP	Dilution	Borat	
1008.5	2865	286.1	302.7	27	13.5	7500
MWe	MWT	°C	°C	m3/h	m3/h	ppm

Date	Power %	#	Comments	Power		Duration/Rate		Temp. °C	D Gr pas	Boron ppm	critic search	Flow Limit	Delta I Band
25/06 13:58	99.8	1	Load initial state	99.8	%		hr	302.2	220	1039.3			
25/06 14:48	50.0	5	power decrease	50.0	%	10	MWe/min	Réf.	DDI	Crit	Bore	Non	0.0
25/06 19:48	50.0	5	plateau	50.0	%	5	hr	Réf.	DDI	Crit	Bore	Non	0.0
25/06 20:38	99.8	5	power increase	99.8	%	10	MWe/min	Réf.	DDI	Crit	Bore	Non	0.0
26/06 01:38	99.8	5	stabilisation	99.8	%	5	hr	Réf.	DDI	Crit	Bore	Non	0.0

Launch calculation	Clean up	Complete	Insert line	Delete line

FIG. V–2. CORMORAN interface (courtesy of C. Schneidesch, Tractebel Engineering, Belgium).

These predictive capabilities use the real core status to generate and analyse several realistic transient scenarios in order to forecast the best feasible plant manoeuvre, leading to the fastest return to full power with a minimum use of effluents. The operator drives the process through an interface (see Fig. V–2), where they specify a scenario and launch the simulation.

(c)   In-core flux map processing

A flux map is measured periodically as per the technical specifications, and the measured data are processed to produce the in-core 3-D power distribution with its power parameters. These parameters are compared with their regulatory limits in order to demonstrate that the core power evolution during cycle depletion remains at all times within the limits that guarantee the safe continued operation of the reactor.

The treatment of the in-core flux maps is an off-line process. A flux map is measured using a set of mobile in-core fission chambers, which provide axial distributions of relative reaction rates in about one third of the assemblies in the core (i.e. the instrumented assemblies). In order to provide a complete power distribution for the whole core, the missing two thirds of reaction rates (i.e. those in the non-instrumented assemblies) have to be reconstructed. This is achieved by a dedicated application that derives the main power distribution parameters, such as the axial offset, the power peaking factor FQ and the quadrant power tilts.

Various techniques have been used to reconstruct a complete 3-D power distribution from in-core measurements, some using empirical correlations to fill the missing measured values. The approach developed in the framework of CORMORAN makes the theoretical core physics calculation match the measured reaction rates, where the 3-D flux solution across the core is the one that fits to the measured data without reconstruction bias via the reaction rates. The principle is to adjust iteratively a best estimate set of calculated nodal 'trial adjustment functions' towards a 'target' that is based on the set of measured reaction rates. By doing so, the original digital core model corresponding to the flux map conditions evolves towards a digital twin of the core. The iterations are stopped when the standard deviation of the differences between the calculated and measured reaction rates has decreased to a value close to the measurement reproducibility error.

Finally, the power distribution produced by the updated core model is used to determine the various safety parameters, such as FQ and fraction of nominal design heat, under calculation conditions that replicate the real core by reproducing measured reaction rates as exactly as possible.

V–3.1.2. Theoretical calibration of ex-core detector signals

Safety functions related to the control and protection of a pressurized water reactor (PWR) in operation are ensured by maintaining a set of relevant core parameters within allowable ranges and measuring them continuously. The in-core power distribution parameters are part of these, but because the design of the Belgian units does not include permanent in-core detectors that would allow continuous access to real time in-core characteristics, these parameters (power, axial offset and quadrant power tilt) are derived through monitoring the ex-core power range detector responses. These detectors capture the neutron flux escaping the reactor vessel, providing an image of the in-core power distribution.

Due to their location outside the reactor vessel, the ex-core power detectors have a limited view of the exact 3-D power distribution in the core. Since most of the information comes from the fuel assemblies at the core periphery, a correlation is needed between the (unseen) in-core distribution and its image monitored by the ex-core signals. The correlation is established periodically by the so-called ex-core calibration process, which consists of perturbing the core by a voluntary axial power swing. During this manoeuvre, in successive flux maps measured over a range of different axial offsets, the in-core flux distribution and the ex-core response are measured simultaneously. The two sets of measurements are compared to calibrate the detectors (see Fig. V–3). The axial core power swing can be performed by either xenon oscillation or control rod movement transient.

As the power distribution of the core evolves during the cycle life, it is important to recalibrate those ex-core power detectors when needed. For each reactor, the correlation should be determined at each new cycle and checked periodically during the cycle life, with technical specifications requiring such transients initially to be performed every three months. These transients are costly in terms of material and human resources and introduce an operational disturbance that is desirable to avoid.

The availability of reliable digital core models allows plant manoeuvres and in-site measurements to be replaced by explicit simulation. Tractebel developed an application to simulate the response of ex-core detectors and a methodology for theoretical ex-core calibration in place of on-site plant manoeuvres. The method relies on PANTHER core simulation, combining flux distribution and importance factors to determine the contribution of each fuel assembly in the core to the detector responses. The assembly importance factors (the probability that a fission neutron born in a certain node will be captured in a given ex-core chamber) are calculated by the 3-D Monte Carlo code MCBEND. They are calculated once for each reactor licensing condition.

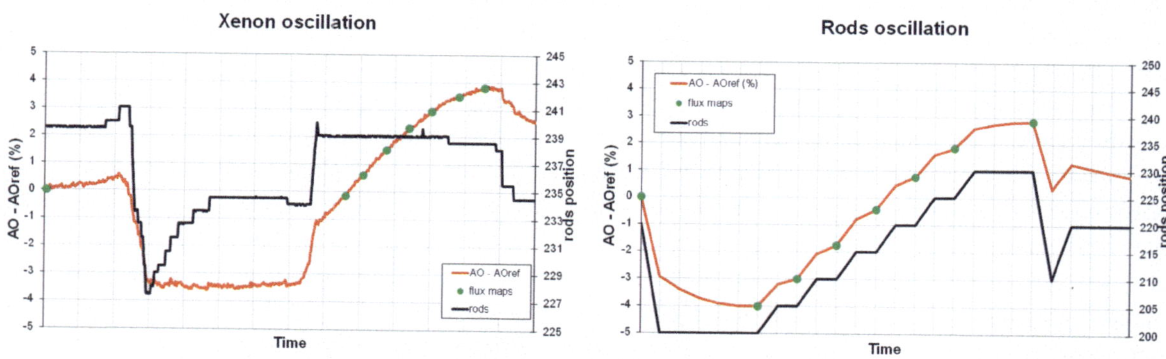

FIG. V–3. Two sets of measurements are compared to calibrate the detectors (courtesy of C. Schneidesch, Tractebel Engineering, Belgium).

Technically finalized in 2005, the method was first approved in 2009 by the Belgian security authorities for progressive application at the start of each new cycle. The feedback demonstrated the high reliability of this theoretical forecast specific to the cycle concerned. The predictive method extension to the entirety of each cycle has been in application since 2010, after approval by the safety authorities and completion by the operator of the upgrade of its operating procedures. The quality and efficiency of a PWR operation is improved by predicting the calibration coefficients by simulation instead of performing core axial swings for calibration transients during startup and in the course of a cycle. Several calibration transients lasting up to 30 hours are avoided for each of the seven nuclear units (up to four per cycle), which translates into direct environmental and economic benefits for the utility: reduced liquid effluents due to dilution and boration (resulting in lower treatment costs), extension of component lifetime through reduced wear of the mobile in-core fission chambers and drive mechanisms, and reduced service requirements.

Correct theoretical precalibration also avoids significant time losses during ramp-up, allowing the full power level to be reached without any delays (up to one day gained). The theoretical approach enables coverage of a wider range of axial offset variation than what certain operational domain restrictions might permit, thereby providing a robust foundation for the calibration process at all times. Early detection of a detector malfunction, identified by an unusual calculation–measurement deviation, facilitates resolution of the issue, avoiding additional studies and tests. The benefit of this approach was demonstrated when an increasing drift in the theoretical calibration was traced to the deterioration of certain ex-core chambers, prompting inspections that led to the replacement of those chambers.

V–3.2. Building information modelling and simulation

Tractebel applies BIM and simulation across multiple aspects of facility management and analysis. This includes enriching computer aided design (CAD) models with additional data layers, developing digital twins to support dismantling projects, performing radiation dose calculations in accordance with the 'as low as reasonably achievable' (ALARA) principle and converting existing BIM models to Modelica models for building energy performance simulations.

V–3.2.1. CAD to BIM: Additional data layers beyond geometries

Tractebel has capabilities in CAD and BIM, as well as in mechanical, electrical and plumbing (MEP) modelling (drafting), from conceptual design to detailed design and site supervision. Tractebel's experience shows that the effective and timely integration of 2-D/3-D models with the conceptual planning and design in a project is crucial in order to achieve the overall project objectives in the best possible time and accuracy at the most cost effective levels by balancing the various project constraints.

Tractebel uses various software tools able to add data from, for example, databases directly during modelling, such as Revit, OpenPlant Modeler and AutoCAD Plant 3D. These data are usually functional characteristics (e.g. design pressure, temperature) or physical characteristics (e.g. material, mass, volume). They are often used for calculating the bill of materials or creating simulation software input. Tractebel uses the Industry Foundation Classes (IFC) format in many developments because it is a very useful and interoperable open format.

Planning and cost data can also be linked to the 3-D/BIM models using software such as Navisworks, Solibri or a dedicated application if desired. These data can be used for visualization purposes (e.g. evolution of a demolition over time) as well as for calculations (e.g. cash flow, waste flows).

Tractebel is also able to add additional layers of information from different types of sources on top of 3-D/BIM models for visualization and calculation purposes. This enhanced visualization allows engineers to make better decisions and identify potential issues more easily. The use of data alongside geometries and the ability to translate them from one format to another also allow setting up calculation models easily and can greatly improve the post-treatment of the calculation results. The automation of data and geometries transfer also reduces the risk of human error while increasing efficiency.

V–3.2.2. A dismantling digital twin

Within the framework of the projects of dismantling the Belgian nuclear power plants, Tractebel is developing a 'dismantling digital twin' with the following functionalities:

— Integration of the plant inventory, consolidating decentralized legacy databases in a common data environment, linked to the digital model;
— Simulation for characterization of contamination and activation;
— Prediction of waste quantities and fluxes per category and optimization of dismantling schedule, cost and resources;
— Design of new waste management facilities and of modifications of existing installations.

To achieve this, Tractebel has developed a common data environment linking the consolidated plant inventory database with the digital model and making them interoperable with different applications. This approach progressively builds an ecosystem of interconnected systems around the digital model, which can be seen as a centralized digital model linked to decentralized applications.

Figure V–4 provides an illustration of the integrated model, where a consolidated plant inventory database is combined with the BIM model. The model can be transferred as an input to simulation software, and the results of the simulations can be reimported and visualized within the model. Figure V–4 shows an example visualization using Monte Carlo N-Particle (MCNP) code to simulate activation of the concrete in the reactor pool.

There is a clear advantage to developing a centralized digital model and using it as an input to simulation software; modelling all systems and structures once, with the most appropriate software tools, provides added value by enabling visualization and post-treatment of the simulation results, thereby improving the level of information in the centralized digital model.

Using the open BIM format (IFC), the project can benefit from typical BIM functionalities: linking the plant inventory elements with the project schedule and cost elements and providing visualization of the nuclear power plant evolution over time. This capability is essential for identifying clashes between new buildings to be constructed and existing structures, as well as for optimizing the planning of various high level dismantling and demolition activities.

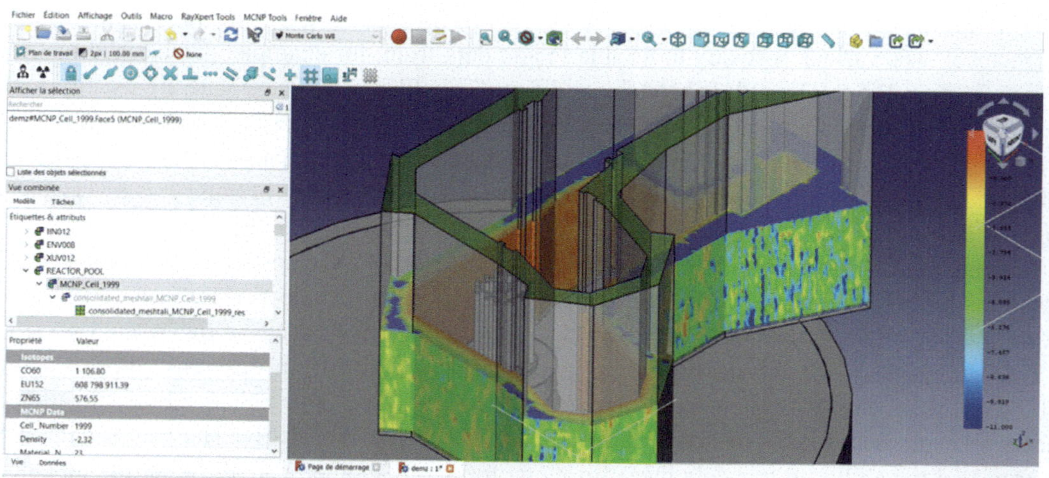

FIG. V–4. Tractebel developments on open BIM strategy (courtesy of C. Schneidesch, Tractebel Engineering, Belgium).

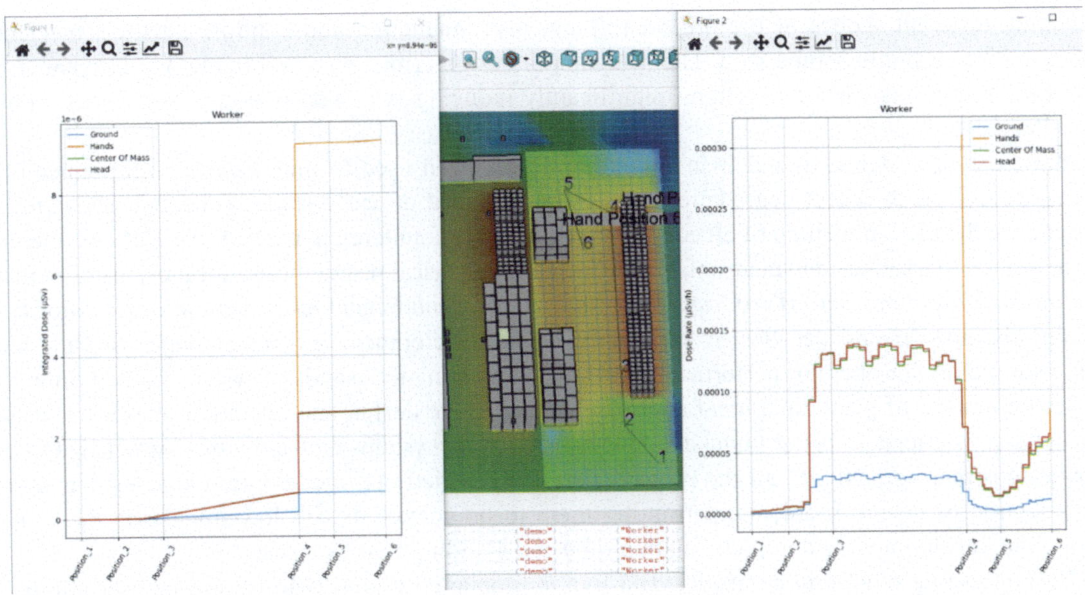

FIG. V–5. BIM coupling to ALARA analyses (courtesy of C. Schneidesch, Tractebel Engineering, Belgium).

V–3.2.3. BIM and ALARA calculations

Tractebel uses RayXpert, a Monte Carlo code, to calculate dose maps in buildings for ALARA analysis. In order to determine individual doses to workers, it is necessary to define their trajectories and time spent at different locations (see Fig. V–5).

A custom tool has been developed to greatly improve the efficiency of this task by loading the RayXpert model and the resulting dose map and allowing the user to select positions directly in the model using mouse clicks and entering the time spent at the different locations selected. Optionally, the hand position can also be specified.

V–3.2.4. BIM and Modelica simulations

Tractebel has built a gateway to convert an existing BIM model to a Modelica model for building energy performance simulations. Modelica is a powerful tool for modelling the physics of complex systems. The developed gateway takes an IFC model of a building that can be generated by Revit, for instance. It takes into account the building walls, floors, ceilings, roof, windows, ventilation system and heat loads and it converts it automatically into a Modelica model.

This enables simulation of temperature evolution in different rooms of the building, taking into account their localization in the building (e.g. north, south), their equipment and heat loads, and their ventilation. Users can also parameterize specific outdoor climatic conditions (not only air temperature, but also radiation, wind speed and cloud cover).

This tool is highly flexible and can also be used to easily implement the conversion from a BIM model to a Modelica model for other types of simulations, such as hydraulic simulations.

V–3.3. System performance simulation

To improve efficiency during the design phase and operation of complex installations, Tractebel uses digital models of the installation to analyse behaviour. During the implementation phase of new installations, simulation of digital twins of the installations has allowed Tractebel to test and evaluate the system even before it was built. This approach allows Tractebel to challenge the design, as well as to integrate different scopes of the project (piping, instrumentation and control, electrics, mechanical works, etc.) into a single comparison. The detection of design errors even before factory acceptance tests are performed permits early corrections, significantly reducing costs and overruns associated with late design changes.

For existing systems, digital twin simulations have been used to monitor the performance of the systems, as well as to detect degradation of the equipment at an early stage. Tractebel has significant experience in detailed modelling of all equipment typically encountered in nuclear installations, including (but not limited to) reactors and neutronics, fluid systems, electrical installations, turbines, instrumentation and control equipment, and HVAC systems, as well as containment environmental conditions. The company also has several decades of experience in the development and maintenance of (full scope) training simulators for operator performance training, simulating thousands of systems in real time.

Using models of plant equipment and systems during the design and operation phases has enabled Tractebel and its clients to better understand new and existing systems and apply this knowledge to build better designs more quickly. In all the projects that were carried out, model based analysis has allowed early detection of design problems, permitting them to be corrected with low impact on the projects, reducing the development and implementation time and avoiding rework costs.

By connecting plant and controller data live to a detailed digital twin of a system or equipment, an in-depth monitoring programme can be deployed, degradations or abnormal plant conditions can be detected early and performance enhancements can be worked out.

V–3.4. Specificities of brownfield/existing assets

Brownfield/existing assets present unique challenges that necessitate precise characterization of both the physical environment and the associated documentation. This includes generating 3-D point clouds of installations, converting legacy drawings into digital formats and creating enhanced 2-D layouts to support safety analyses and operational planning.

V–3.4.1. Point cloud/laser scanning of complex facilities

In existing installations, laser scanning allows analysis of an environment and automatic measurement of a large number of points on the external surface of existing objects to create a point cloud. This point cloud can then be used for measurements, visualization, animation, rendering and construction of a 3-D CAD model of existing installations. Such modelling is especially important for minimizing risk when installing new equipment in an existing facility.

In addition, 3-D scans allow virtual walkdowns of the installation, which are useful for inventorying equipment within the installation and reduce the need for multiple on-site walkdowns, resulting in substantial efficiency gains.

Tractebel has considerable experience in acquiring point clouds, as well as cleaning, aligning, segmenting and using them for 3-D modelling purposes. The company has conducted large 3-D laser scan measurement campaigns of the complete reactor buildings of Tihange 3 and Doel 4 (all levels), comprising more than 500 individual scans for Tihange 3 and 800 for Doel 4. The point clouds from each measurement campaign were assembled into single comprehensive files (one for each reactor building).

V–3.4.2. Two dimensional legacy drawing digitalization: Vizer

Recent advances in computer vision and deep learning have allowed the development of new computer techniques to automatically recognize objects in engineering drawings. These objects can represent equipment, lines or text.

Such capabilities pave the way for a 'paper to simulator' workflow: starting from a plain engineering schematic, it becomes possible to identify all entities (equipment, lines, metadata such as text, etc.) and then infer the relationships between them (e.g. metadata to equipment, lines to equipment, equipment to equipment connected by lines, drawing to drawing). By recognizing metadata found on the drawing, it is often possible to query additional information through other available databases.

Within Tractebel, work has been performed to leverage those techniques in order to design an application allowing engineers to input an engineering schematic without any data and obtain as an output a digitized version of the drawing, where all the information (from the drawing, from additional databases) about the drawing is gathered in one verified place. Furthermore, the application provides additional capabilities, such as highlighting equipment in specific colours according to safety criteria and exporting collected data and user added data to tabular formats (e.g. an Excel file). The application improves the efficiency of the engineers working with such schematics, as well as the quality, through bullet-proofing and better collaboration tools. The application developed has focused on digitizing piping and instrumentation diagrams.

V–3.4.3. Two dimensional enhanced maps: Safety Reports

For fire hazard analysis studies and other purposes, it is important to have comprehensive information about the layout of the installation (size and shape of rooms, relative positions, existing openings, fire compartment borders, etc.), an inventory of the contents of the installation/rooms (i.e. the fire load, fire detection and suppression systems, safety equipment, cable routeing, etc.) and their characteristics (dimensions, materials, loads, etc.). Unfortunately, when such information exists, it is often dispersed across multiple data sources and is rarely detailed enough for the specific purpose of fire hazard analysis studies.

A central repository is therefore needed to consolidate all these data and provide capabilities for integrating specific calculation modules that can then use the data to simulate scenarios of fire behaviour and propagation, for example. This is particularly challenging with older units that were not designed in the digital era. A simple and effective solution is needed.

For this purpose, Tractebel uses the Safety Reports software developed by Certeso. This tool is comparable to a BIM tool, except that the view is 2.5D. All the 2-D layouts of the installation at different elevations are loaded into the application; the rooms are then identified and detailed (structure of the walls, openings, etc.), and the contents are detailed by positioning the various relevant pieces of equipment and characterizing them.

A specific calculation module has been developed by Tractebel and integrated into the tool. It uses the data to calculate the information for the study under different scenarios (room where the fire starts, ventilation on or off, doors open or closed, etc.) to evaluate the potential impact on the safe operation of the plant.

The Safety Reports software can work in coordination with the Safety Walk application, which runs on Android tablets and facilitates on-site inspections and data collection in order to complete the information in the central system. The application can be linked via Bluetooth to a laser meter to take measurements and register them automatically on a tablet, or to a 360° digital camera. Other photographs can be taken using the tablet itself.

All collected information is transferred and made available in the Safety Reports software.

V–3.5. Virtual and augmented reality

VR and AR technologies provide immersive and interactive ways to visualize and assess project models throughout their development.

V–3.5.1. Virtual reality applications

Tractebel has implemented a VR solution for its projects comprising a dedicated VR room equipped with a video projector and VR headsets. This set-up enables full immersion in the 3-D BIM models of

FIG. V–6. Tractebel AR solution on-site (real view and augmented view through Microsoft HoloLens headset) (courtesy of C. Schneidesch, Tractebel Engineering, Belgium).

projects, allowing users to navigate future installations, ensure overall consistency and optimize the design and implementation of projects.

From the design phase onwards, Tractebel's VR solution allows users to experience a project at human scale, move through the installations, take measurements and photographs and reposition objects while interacting with collaborators around the model. This VR solution fits perfectly into the BIM processes and requires only exported files to generate VR renderings.

V–3.5.2. Augmented reality applications

Tractebel has identified the added value of AR for buildings' site follow-ups, technical reviews and on-site access to additional data sources. AR technology allows users to access large amounts of information and connect with experts all over the world while on-site. Tractebel chose commercial hardware and software solutions (see Fig. V–6). Key features of this AR solution include:

— Faithful immersion in 3-D models in situ;
— Continuous, accurate alignment of the 3-D model and the existing environment;
— Real time annotation, usable in a back office alongside the 3-D model;
— Collaboration;
— Pilot–passenger mode to guide users unfamiliar with AR.

V–3.6. Systems engineering

The International Council on Systems Engineering (INCOSE) [V–2] defines systems engineering as "a transdisciplinary and integrative approach to enable the successful realization, use, and retirement of engineered systems, using systems principles and concepts, and scientific, technological, and management methods." Systems engineering is key in nuclear projects, especially in nuclear new build projects. Tractebel has access to a variety of commercial and in-house tools to perform systems engineering activities.

At Tractebel, the systems engineering capabilities are linked with BIM and digital twin technologies. In practice, the systems engineering processes of traceability and verification of change to engineering and project data are linked with the BIM and the digital twin through their backbone (i.e. the product breakdown structure). The philosophy behind this development is to link the systems engineering functionalities with the BIM and digital twin functionalities to create an integrated, connected database for both digital twin and systems engineering that supports:

— Requirement management, including assumptions and inputs;
— Interface management (between studies; documents and deliverables; structures, systems and components; contractual lots);
— Configuration management and design change processes (project data management, design choices, design modifications, bill of materials/bill of quantities);
— Engineering management workflows (task assignment, action lists, deliverable lists);
— Layout and 2-D/3-D/BIM models.

The filtering capabilities are implemented with improved visualization tools, including tree views (breakdown structures) and relationship diagrams (interfaces) with arrows, complementing the table views and providing a user friendly interface.

REFERENCES TO ANNEX V

[V–1]   NUCLEAR REGULATORY COMMISSION, The State of Technology of Application of Digital Twins, US NRC TLR/RES-DE-REB-202101, NRC, Washington, DC (2021).
[V–2]   INTERNATIONAL COUNCIL ON SYSTEMS ENGINEERING, INCOSE Systems Engineering Handbook, 5th edn, Wiley, Hoboken, NJ (2023).

Annex VI

HOW DIGITAL TOOLS CAN SUPPORT THE EFFECTIVENESS OF ENGINEERING PROCESSES IN NUCLEAR POWER PLANTS

VI–1. INTRODUCTION

The nuclear division of CEZ Group has been implementing its digitalization programme since 2017, through several projects and initiatives. Progress has been achieved gradually and with an agile approach, resulting in changes and improvements to software tools as well as to organizational behaviour. In some cases, mobile devices have been implemented as a replacement for paper based agendas. The contribution of knowledgeable personnel is essential to the success of digitalization projects. It is important to have a combination of knowledge: an expert group should understand the processes while also having an overview of available software tools and options.

VI–2. PAPERLESS APPROACH

Several paperless solutions have been implemented through software applications. For example, PowerFlow has been implemented in the nuclear division, with plans to extend its use to other divisions of CEZ Group. Five agendas have been converted from paper to PowerFlow, and a legacy application has been migrated to a new paperless platform as well. This transition eliminates the need for paper and improves work efficiency (see Fig. VI–1).

PowerFlow is a workflow engine in which a designer can set an agenda that will have an exact process, data inputs and outputs. The software is owned by the PowerFlow company and is based on the Camunda platform. Additional components include a template engine, digital signature capability and other mobile support.

Mobile support has been available in the nuclear division since 2018. CEZ started with support for the tagout process and later expanded mobile support to include support of work orders for team leaders in work conduct.

Two platforms are used: a web based platform for general administration and tagout and work order coordination, and an Android based platform on mobile devices for the tagout process and work conduct. The project has been implemented using an agile approach, with close attention to the needs of the end

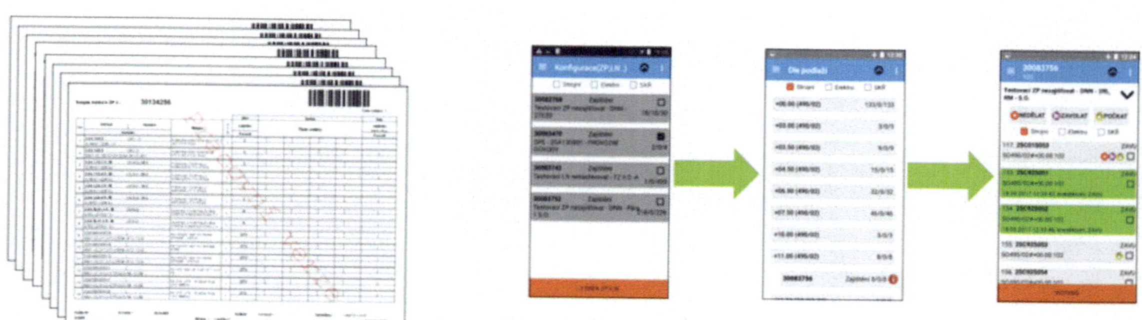

FIG. VI–1. *Tagout order transformation from paper to mobile device (courtesy of P. Nekula, CEZ Group, Czech Republic).*

users for operation, radiation protection and system engineering, as well as contractors. Wi-Fi hot spots have been installed in specific areas of the plant, and mobile devices are synchronized manually when within range of a Wi-Fi signal.

VI–3. VIRTUAL REALITY AND 3-D

Different levels of 3-D scanning and modelling are used in CEZ Group nuclear plants. A comprehensive 3-D model is mainly accessible for the Temelín nuclear power plant, while 3-D scans are available at the Dukovany nuclear power plant. These 3-D scans and models are available within an application that provides users with direct access to all the information included: plant situation, levels of equipment and space around equipment. This capability reduces the need for on-site visits, saving time for other activities. Virtual reality rooms have been set up at both plants to support training for specific roles during activity preparation, using immersive virtual reality tools (see Figs VI–2 and VI–3).

VI–4. OTHER TOOLS AND TIPS FOR EFFECTIVENESS IMPROVEMENT

Process mining techniques are used for faster information about analysed processes (see Fig. VI–4). Business intelligence plugins are also used within the nuclear division.

A robotic process automation (RPA) approach has been used in the nuclear division for mass activities related to repeated downloads of documents from a dedicated web page. The RPA system is configured to log in to a web page, open a specific link and save a document from a dedicated link. Additional steps are under review. The objective is to reduce time spent on repeated activities. CEZ Group uses the UiPath solution for this purpose.

FIG. VI–2. A 3-D scan of the main circulation pumps deck (courtesy of P. Nekula, CEZ Group, Czech Republic).

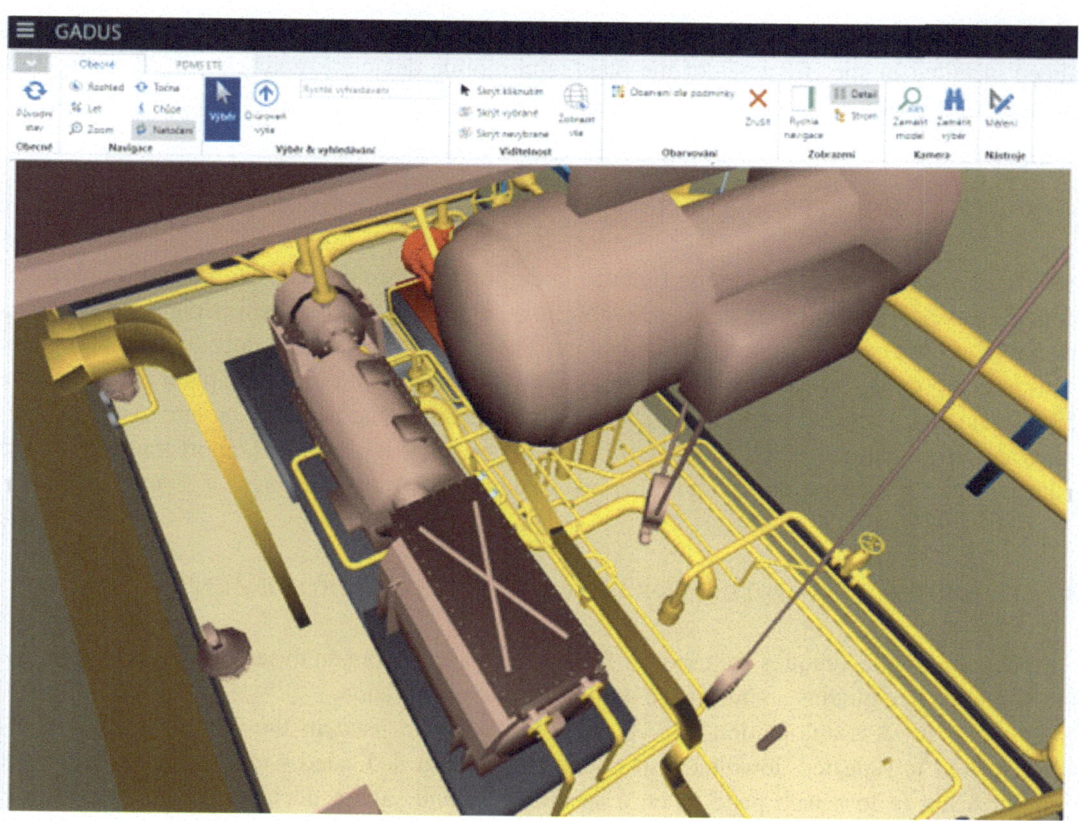

FIG. VI–3. Example of a 3-D model in the turbine hall (courtesy of P. Nekula, CEZ Group, Czech Republic).

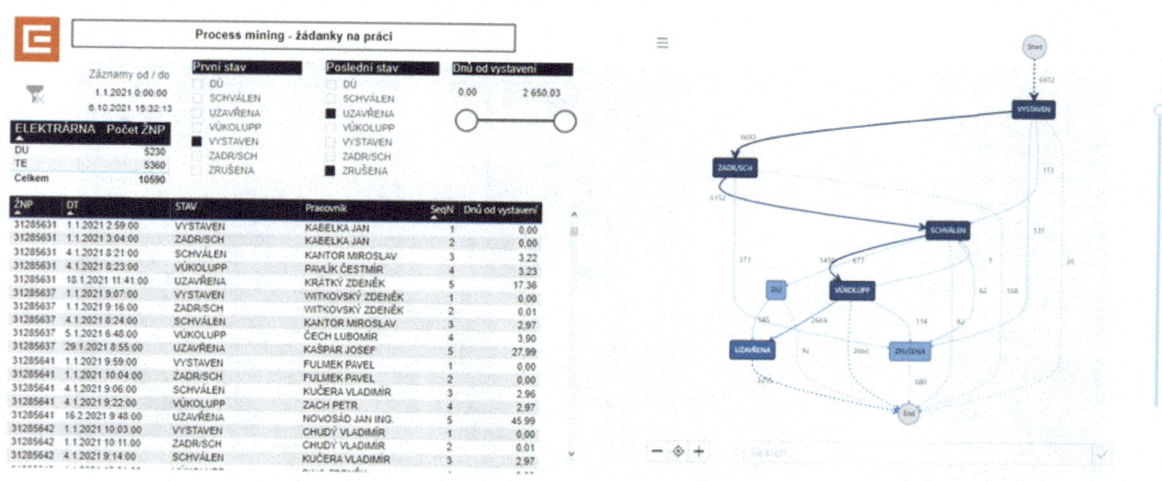

FIG. VI–4. Process mining example for work order requests (courtesy of P. Nekula, CEZ Group, Czech Republic).

INTEGRATED VIRTUAL STIMULATION OF THE APR1400 MAN–MACHINE INTERFACE SYSTEM FOR HIGH AVAILABILITY AND OPTIMIZED OPERATION

VII–1. INTRODUCTION

Stable electricity production is an important performance indicator for nuclear power plants. However, sudden scram events can occur and are difficult to prevent, as they may arise from various causes. In the Republic of Korea, issues with the reliability of digital instrumentation and control (I&C) systems and human error are widely recognized as major contributors to scram events. The information presented in this annex concerns improving the reliability of digital I&C systems and reducing human error in nuclear power plants.

Korea Hydro & Nuclear Power (KHNP) operates two APR1400 reactor units and is constructing four additional units in the Republic of Korea. The localization of the APR1400 Man–Machine Interface System (MMIS) refers to digital I&C systems, including the main control room, installed in Shin Hanul Units 1 and 2 and Shin Kori Units 5 and 6. The localized APR1400 MMIS was successfully developed as a first of a kind system in 2012, but it still lacks dedicated simulation software. The emulation or stimulation typically offers several advantages [VII–1]. Test distributed control system (DCS) or programmable logic controller (PLC) control strategies can result in a pretuned control system. Before replacing a DCS, potential input/output (I/O) issues with the new DCS can be identified, which can help make the control algorithm of the early development stage more accurate [VII–2]. For Shin Kori Units 3 and 4, the MMIS, manufactured by Westinghouse, has been installed and is operating now. A special simulation function, based on the 'virtual Ovation' DCS controller, has been partially adopted in the Shin Kori Units 3 and 4 operator training simulator. This tool can be classified as 'emulation', covering only some parts of non-safety I&C systems; other parts depend on the simple simulation approach.

Virtualization technology has been widely used in many kinds of applications. For example, VMware software enables an x86 Xeon server to be partitioned into multiple small virtual computing machines, each with one or more central processing unit (CPU) cores, designated memory and storage. This approach is called server virtualization. Another virtualization technology is hardware platform emulation, such as using a hypervisor. The hypervisor does not emulate the CPU itself, but only the instruction sets of the CPU. The hardware platform includes the CPU and the associated peripherals, such as the mainboard bus, PS/2 keyboard and mouse, serial port and floppy disk. Based on this emulated hardware, it is possible to run a specific real time operating system (RTOS) and its applications.

The APR1400 MMIS uses two digital platforms: POSAFE-Q for safety systems, based on a safety grade RTOS, and OPERA-1400 for non-safety systems, based on another RTOS, called VxWorks. VxWorks is designed for use in embedded systems requiring real time, deterministic performance and, in many cases, safety and security certification, for industries such as aerospace and defence, medical devices, industrial equipment, robotics, energy, transportation, network infrastructure, automotive and consumer electronics.

In 2020, KHNP launched a research and development (R&D) project to develop the APR1400 MMIS virtualization simulator, known as 'virtualized MMIS', to support various engineering activities [VII–3]. This annex describes how the virtualized MMIS can contribute to plant stability and cost optimization from the design and construction phase through to the operation and maintenance phase. Figure VII–1 illustrates potential applications of the virtualized MMIS. Unlike the virtual controller mentioned earlier, the virtualized MMIS can handle both safety and non-safety I&C systems, except for

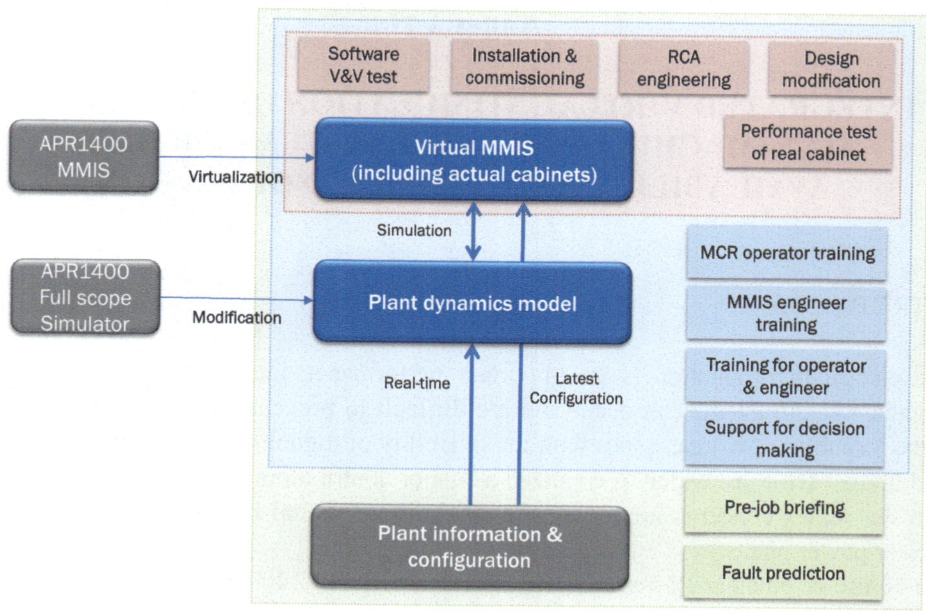

FIG. VII–1. Overview of the virtual stimulation of the APR1400 MMIS. MCR — main control room; RCA — root cause analysis; V&V — verification and validation. (Courtesy of S. Lee, Korea Hydro & Nuclear Power Company, Republic of Korea.)

certain platforms that are mainly used for specialized monitoring of the power plant, such as for seismic, fire and external events.

Global vendors such as Siemens, Emerson, ABB and Framatome provide virtual controllers (see Ref. [VII–1]). These can be used for plant logic verification, I&C engineer and plant operator training, and plant logic development. However, it is difficult to find a simulation tool based on the virtual controller concept that provides the ability to describe the network or connection architecture of a real plant's digital I&C systems. The KHNP R&D project aims to develop a virtualized MMIS and associated plant dynamic models for versatile use throughout the overall life cycle of a nuclear power plant. The virtualized MMIS can simulate PLC and DCS platforms down to the level of main components, such as processor boards, I/O boards and communication boards, as well as various interconnections among the platforms and computing devices. It also includes the operator human–machine interface (HMI) and the main control room's various user interfaces. The virtualized MMIS cannot be classified using existing classification criteria; therefore, a new category is proposed: 'virtual stimulation'. This approach implements the actual software and identical hardware configurations for PLC and DCS systems, making it fundamentally different from the existing PLC or DCS simulation method.

VII–2. DEVELOPMENT OF A VIRTUALIZED MAN–MACHINE INTERFACE SYSTEM

The virtualized MMIS can be regarded as the MMIS digital twin. A 'digital twin' is a technology that predicts outcomes by creating replicas of real world objects on a computer and simulating situations that may occur in the real world. The term is used for a wide variety of purposes, and because of the recent popularity of this term, it may evoke preconceived notions before the development of the virtualized MMIS is explained. All software operating in the real digital I&C system used in nuclear power plants will run in the virtualized MMIS through 'digital duplication', implemented as virtualization, without any additional modification. This approach allows the scope of verification and validation of a digital I&C system to extend beyond its control algorithm, which helps to ensure that the digital I&C system has high reliability.

Replicating the configuration and operation of the real systems is expected to be very helpful in improving business efficiency and system reliability in terms of equipment production by vendors and plant operation by utilities. This approach can directly contribute to securing design quality, shortening the production and commissioning test periods through virtualized MMIS based preliminary function verification and testing in the design/production/test stage. Even in the operation stage, cause analysis based on the virtualized MMIS allows a problem to be reproduced to help identify the cause. In addition, possible errors can be detected in advance by performing the preliminary verification through the virtualized MMIS when design changes or function improvements are made. The virtualized MMIS also helps to improve the education and training of I&C maintenance engineers and control room operators associated with the digital I&C system. In particular, when they perform the collaborative task for the periodic test of safety system facilities, the virtualized MMIS provides a virtual environment that mirrors the actual field conditions at each location. This reduces human error and strengthens employees' engineering competency. Although the virtualized MMIS can be used alone, it should be integrated with the APR1400 dynamic simulation model to support the various applications mentioned above. This model is the basis for simulating various nuclear power plant states. All the MMIS parts in the existing APR1400 full scope simulator are transformed to be replaced with the virtualized MMIS. For field use, it should also be possible to simulate based on the current or real time state of the nuclear power plant. The virtualized MMIS also interfaces with the physical I&C cabinet. A part of the MMIS systems can be replaced with the real cabinets, and other parts are implemented as virtualized cabinets. This makes it possible to check whether the requirements are met and whether the system operates stably in the integrated environment, by linking some physical cabinets to the virtual MMIS as part of the performance verification of the real facility or the improvement of substitutes or facilities. The configuration of the virtualized MMIS varies depending on the physical facility.

VII–2.1. Comparison with the existing training simulator

In the case of a general operator training simulator, the digital I&C system implements the control logic for each system in software from a functional perspective without considering the unique characteristics of the I&C equipment used in the nuclear power plant. In other words, the digital I&C system of the nuclear power plant is not installed as it is (see Table VII–1). The purpose of a general operator training simulator is to replicate the main control room faithfully from the operator's point of view. Consequently, there are limitations in terms of function and performance for the use of MMIS engineers.

VII–2.2. Development of the virtualized Man–Machine Interface System

Figure VII–2 shows the overall APR1400 MMIS architecture for Shin Hanul Unit 1. The red dashed box denotes the virtualization scope. Several monitoring systems are excluded from this implementation. However, there are about 250 cabinets and their 470 racks to be virtualized in this R&D project. The virtualized MMIS development is mainly composed of MMIS virtualization, various engineering applications, several real I&C cabinets and plant dynamic models. The MMIS virtualization is divided into the development of virtualization core technology and the creation of a virtualized system by integrating the products of the core technology. Virtualization core technologies include virtual PLCs, virtual DCSs, virtual I/Os and virtual networks. Engineering applications encompass MMIS engineering functions for analysis, simulation and testing, as well as functions for managing the environment of the virtualized MMIS. For example, one MMIS engineering function is to display the real time dynamic appearance of all I&C cabinets within the virtualization scope based on Shin Hanul Unit 1 and to operate the equipment in real time according to the engineer's manual controls and operations. An important characteristic of the MMIS virtualization is the creation of a virtualized MMIS for a specific nuclear power plant of APR1400 using a software defined architecture approach on a high performance computing machine. The software defined architecture is implemented as one of the exclusive engineering tools referred to as 'smart engineering tools', enabling the creation of various MMISs.

TABLE VII–1. IMPLEMENTATION FIDELITY OF THE VIRTUAL STIMULATION OF MMIS

Aspect	General training simulator	Virtualized MMIS
Logic implementation	Develops control logic diagrams by simulation or emulation	Deploys the real plant logics using the associated engineering tool
Controller hardware/ software	Not applicable	Emulates the controller hardware and deploys the real controller software
Controller I/O	Not applicable, but simulates processing variables	Emulates I/O modules and simulates the associated variables
Connections	Not applicable	Implements the network and connection topology
Operator interface	Implements with emulation or simulation	Deploys the real plant's HMI software
Engineer interface	Not applicable	Develops the required visualization of all the MMIS cabinets
Plant model	Satisfies ANSI/ANS 3.5 criteria	Uses the model of the training simulator

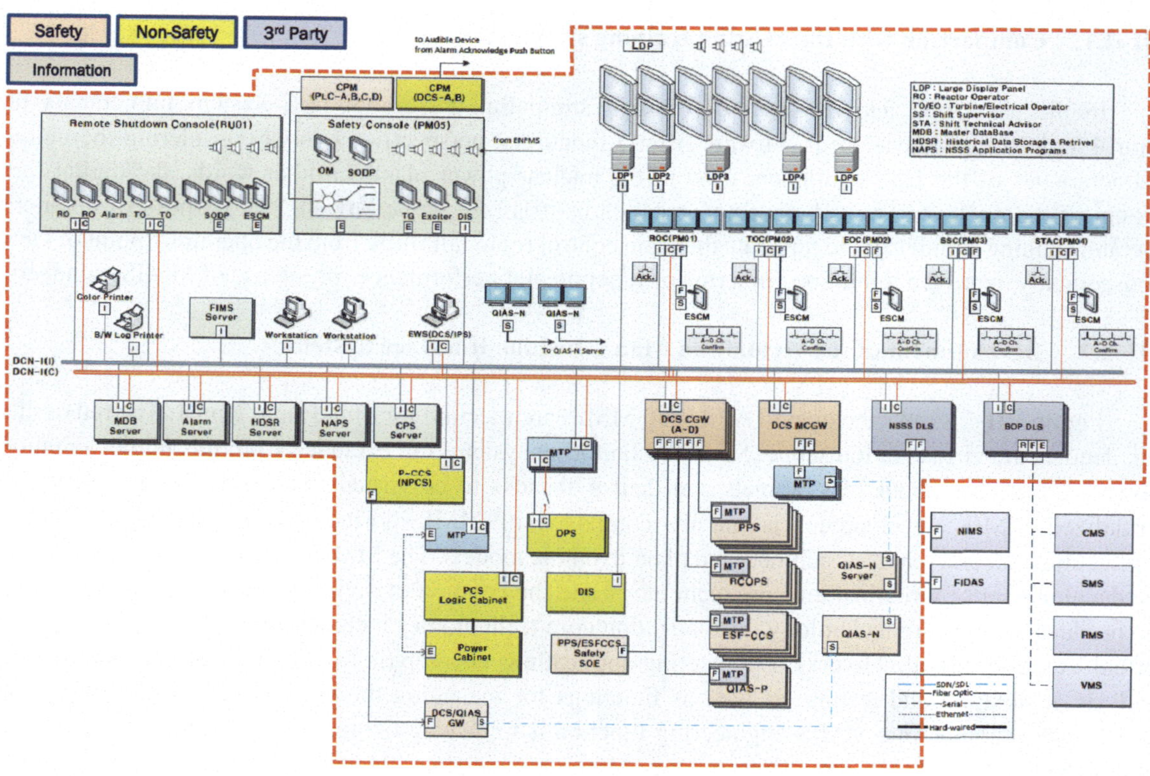

FIG. VII–2. Virtualization scope of the APR1400 MMIS (courtesy of S. Lee, Korea Hydro & Nuclear Power Company, Republic of Korea).

Figure VII–3 shows how an actual MMIS can be divided into unit elements to be virtualized (left panel) and, conversely, how a virtual MMIS can be assembled from those elements according to its hierarchy (right panel). Disassembly and assembly are fundamental concepts in virtualization. The virtualized MMIS has been implemented with various levels of virtualization technologies, from field sensors to main control room software systems. Appropriate virtualization techniques have been identified for each area of the MMIS, including:

— Server virtualization for the HMIs and displays in the main control room;
— Network virtualization based on the hypervisor function for the network and connection topology of all the MMISs;
— CPU emulation to replicate the behaviour of the APR1400 PLC and DCS process modules;
— Software implementation for I/O cards, terminal blocks and other such hardwired functions;
— Plant dynamic models for the sensors, actuators and field components (including the reactor).

VII–2.2.1. Virtual PLC and DCS

One of the most important parts in the development of the virtualized MMIS is virtualizing the CPUs and their associated subcomponents on the processor modules of the virtual PLC and DCS. This is particularly challenging because the target of CPU virtualization is an industrial embedded system rather than a general computer. Virtualization technology that implements the operation of a CPU used in a computer or embedded system as software is currently used in semiconductor manufacturing, automotive electronic control unit development and CPU based home appliance development. Figure VII–4 shows the virtual PLC and DCS being developed in this project. The first step of developing a virtual PLC or DCS is to identify the processor specifications. The next step is to determine the level and scope of virtualization of the processor board.

The final step is to implement the external interface part associated with the virtual control module and the status monitoring of the virtual control module. The virtual PLC has been developed in this way.

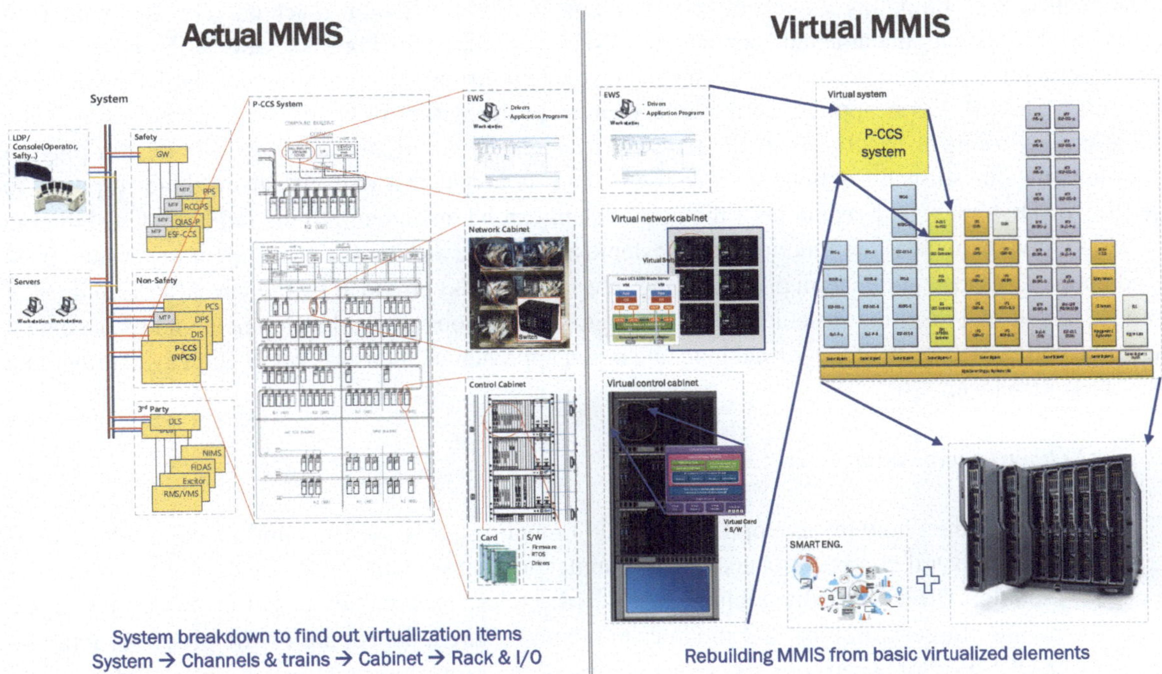

FIG. VII–3. Implementation concept for virtualization of I&C systems (courtesy of S. Lee, Korea Hydro & Nuclear Power Company, Republic of Korea).

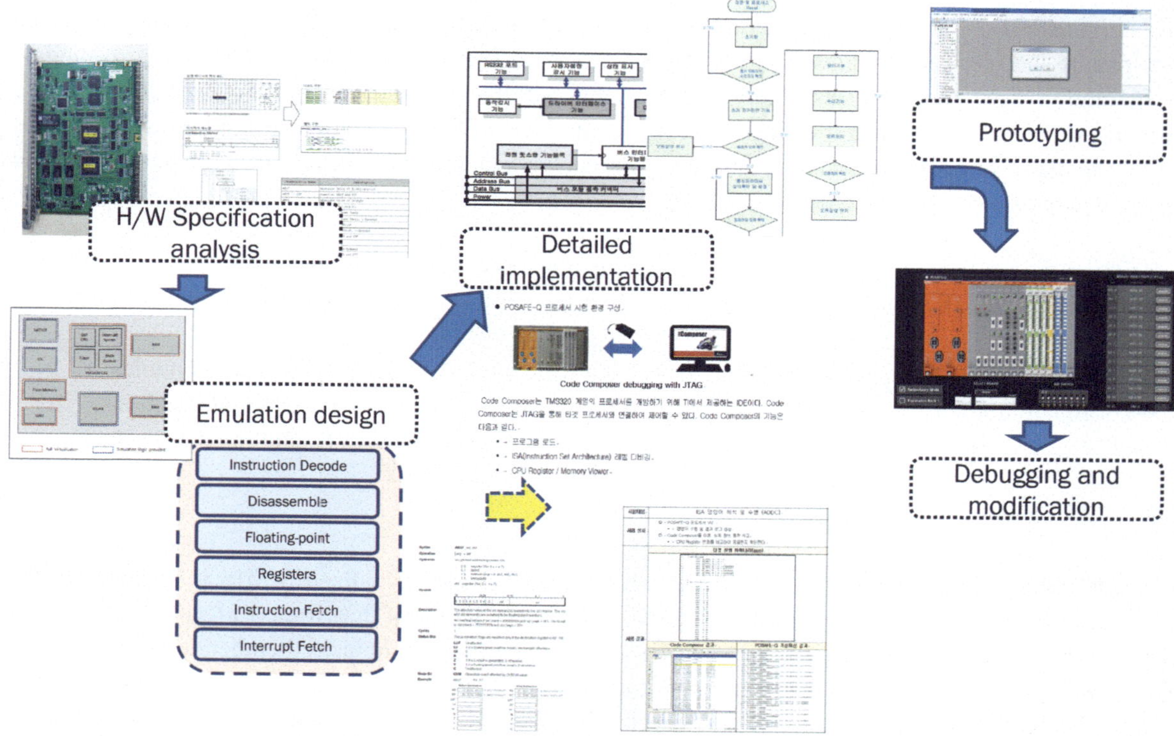

FIG. VII–4. Development process of a virtual PLC. H/W — hardware. (Courtesy of S. Lee, Korea Hydro & Nuclear Power Company, Republic of Korea.)

Task applications and control logic can be created and uploaded through the dedicated engineering tool (pSET-II), just as with the physical PLC. The same applies to the virtual DCS. The developed virtual PLC and DCS modules should be operated in a general computing environment. To ensure hardware independence, the computing environment is implemented with a virtual machine. The deployment of virtual machines is a fundamental element of MMIS virtualization. For PLC and DCS modules, the internal structure can be quite complex. This is the standard virtual machine structure. Full virtualization of the PLC/DCS RTOS is applied, so the binary codes of the virtual PLC and DCS modules run in the emulation environment exactly as in the physical systems. This virtual machine structure can play an important role for software defined architecture. There is a rule that one virtual machine responds to one PLC or DCS rack. The real data of the power plant can be uploaded into the red box area shown in Fig. VII–5, and many possible functions can be deployed into the area on the guest operating system (OS) based on the dedicated application programming interfaces (APIs) provided by the virtual PLC and DCS. Because this virtual machine structure can be configured in various ways, development and management are simplified. Advantages of the virtual machine based approach include scalability, high availability and ease of maintenance.

VII–2.2.2. Other virtualization cores

Before explaining how to develop the virtual I/O functions, it is useful to outline the mechanism of the real product. Several components enable communication between the field device and the controller processor module. These typically include hardwired cables, terminal blocks, I/O modules and the processor module. In general, most I/O modules are manufactured, depending on the processor module, by a PLC or DCS vendor. Therefore, the virtualization of the processor module and I/O modules should be developed by the same vendor, whereas the other components can be independent from the vendor. The development of the virtual I/O can be divided into two components: the inside and the outside of the virtual PLC. The inside component comprises the I/O modules. Several types of I/O modules have

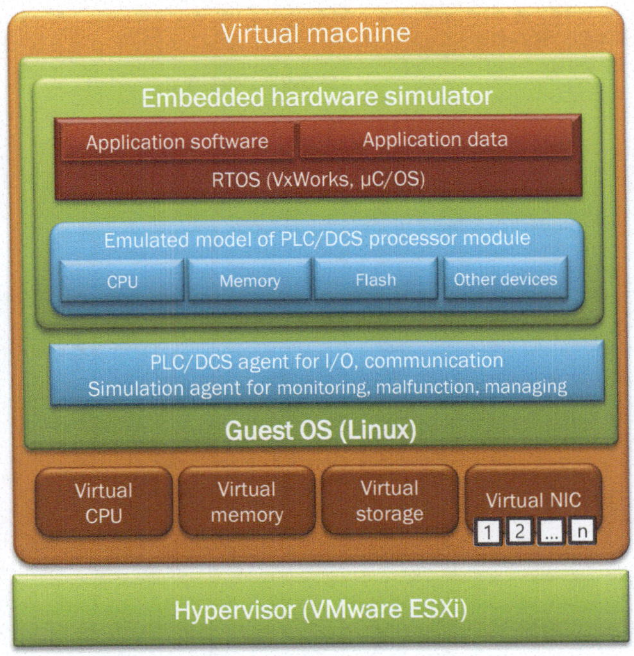

FIG. VII–5. Structure of a virtual machine for virtual PLC and DCS modules. NIC — network interface card. (Courtesy of S. Lee, Korea Hydro & Nuclear Power Company, Republic of Korea.)

been developed. In the real product, these modules are assembled in the form of one rack; similarly, the virtualized modules can be assembled by assigning an empty slot with a specific card in a virtual machine. Numerous possible module configurations can be implemented within the virtual PLC and DCS. The outside component includes cables, terminal blocks and so on, which transfer information between elements. If the emulation of those components is not necessary, a simulation approach is sufficient to implement their virtualization. In the APR1400 MMIS project, this has been implemented as an API within a virtual machine close to the virtual PLC core. External software can use the API to access the virtualized I/O modules. The external software reads and writes the values of multiple virtual PLCs through the API. This is the functionality of the virtual I/O.

The APR1400 MMIS has two main kinds of communication systems: Profibus for the safety I&C systems and Ethernet for the non-safety I&C and information processing systems. Profibus has been implemented using standard serial cables. Given the characteristics of the virtualization, the Ethernet based communication systems are directly deployed in the virtual environment, whereas the Profibus based communication systems need to be converted to Ethernet. Like the virtual I/O implementation, for the non-Ethernet based communications, the networks should be virtualized. Therefore, the information of the safety data network (SDN) or safety data link (SDL) modules of the real PLC should be converted to Ethernet based data and then transferred to external software. These communication types are a kind of network interface method. In general, communication systems operate on a network topology. The APR1400 MMIS virtual network deals with the network interface methods and the communication system's topologies for the digital I&C systems. If a physical network switch is fully virtualized, network virtualization for Ethernet based systems is straightforward. If the switch is not virtualized, alternative means need to be considered. The virtual switch of the hypervisor has been used to achieve network virtualization, and the Profibus networks have been converted into separate Ethernet networks. In order to integrate the virtual PLC with the real PLC, a special conversion program called SSGW (SDN/SDL Gateway) has been developed (see Fig. VII–6). SSGW has been used for the validation and verification of a specific safety system of Shin Kori Units 5 and 6 (see Fig. VII–7).

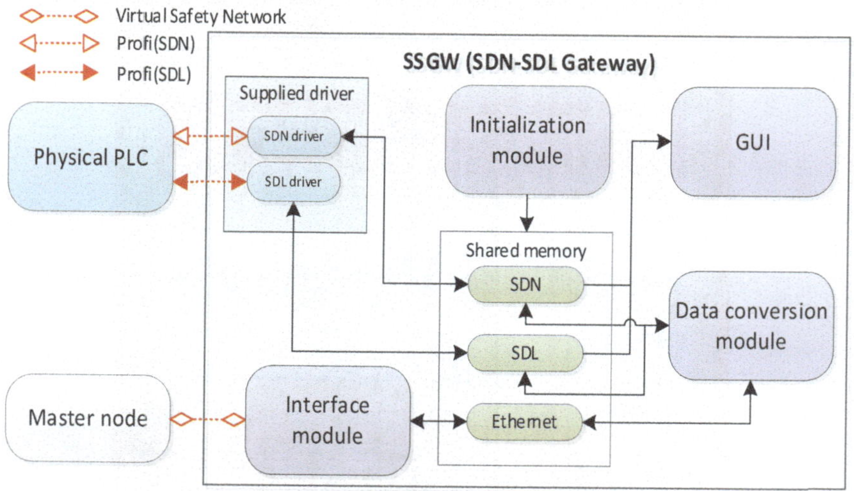

FIG. VII–6. Simple architecture of SSGW. GUI — graphical user interface. (Courtesy of S. Lee, Korea Hydro & Nuclear Power Company, Republic of Korea.)

FIG. VII–7. Integration between the virtual safety system and the physical safety system using SSGW. ITPP — input termination patch panel; ITPS — input termination patch shelf; PC — personal computer. (Courtesy of S. Lee, Korea Hydro & Nuclear Power Company, Republic of Korea.)

VII–2.2.3. MMIS virtualization

For the MMIS virtualization, one virtual machine corresponds to one digital I&C cabinet. In the case of Shin Hanul Unit 1, there are about 250 cabinets. To virtualize all of them, hundreds of virtual machines need to be created, and engineering work is needed to replicate the configuration of the real plant. To use the virtualized MMIS in the power plant office, it is useful to deploy it not in the form of large server clusters but rather as a high density integrated blade server. In Fig. VII–8, each vertical column represents one blade, and each block represents sub-I&C systems composed of multiple virtual machines. A blade server consists of multiple blades. One of the advantages of the blade server format is that several blades can be added if there is not sufficient capacity. For detailed deployment, the performance load of individual blades and virtual network connections needs to be taken into account.

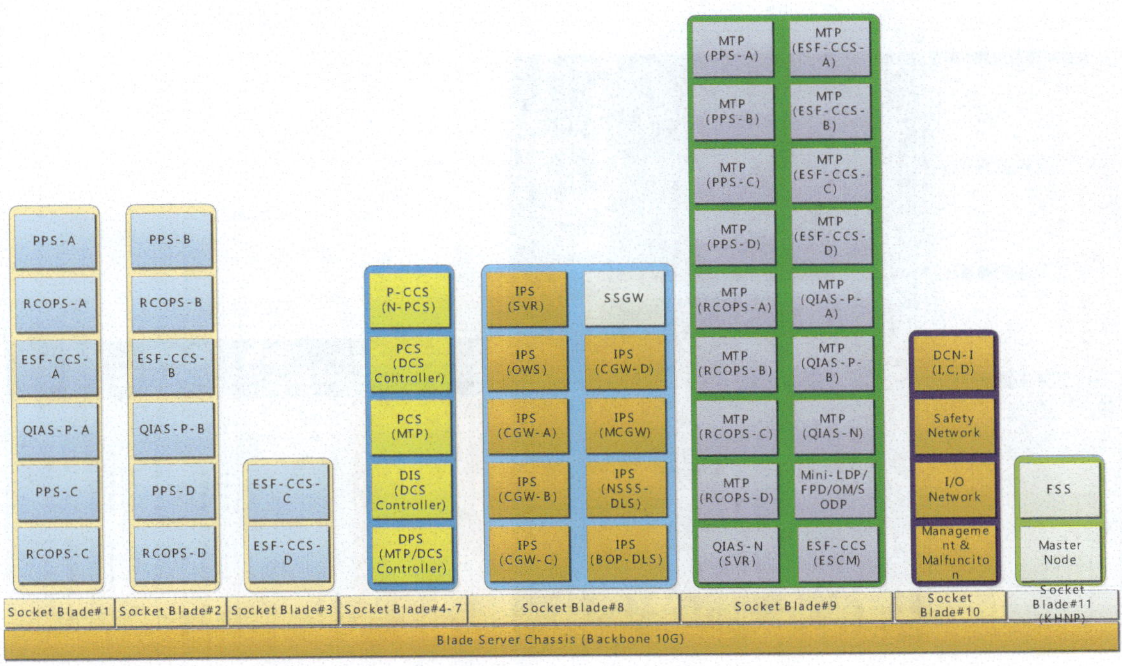

FIG. VII–8. *Typical deployment of virtualized I&C systems on a blade server. (Courtesy of S. Lee, Korea Hydro & Nuclear Power Company, Republic of Korea.)*

VII–3. APPLICATIONS OF THE MAN–MACHINE INTERFACE SYSTEM VIRTUALIZATION SIMULATOR

The real time embedded OS, firmware, application software and engineering tools of the real PLC or DCS run identically in the virtualized MMIS. In addition, the network topology and non-software-based connections, such as I/O, hardwired cables and safety Profibus networks, are also virtualized. A specific engineering tool for the virtualized MMIS, referred to earlier as a 'smart engineering tool', has been developed in order to perform tasks such as building a virtual I&C cabinet using the virtual PLC and DCS, adjusting the virtual cabinet, creating a test scenario with malfunctions and displaying the network monitoring results and the real time internal and external status of all the virtual cabinets. This enables the various tests, analyses and engineering activities to be carried out within the virtualized MMIS, including some that are difficult to perform in the real plant. For example, obtaining the buffer memory status of a virtual PLC in a real world set-up is challenging, but within a virtual PLC environment, it is straightforward to visualize the real time status of internal components.

VII–3.1. Design and construction phase support

Figure VII–9 shows the example of a virtual cabinet with several PLC racks. The smart engineering tool enables the cabinet and rack editors to modify the assembled structure or its subcomponents. This function supports the quick prototype development of new cabinet and rack designs, and the structure of a specific I&C system can be changed. For example, if a real I&C system has five racks and multiple I/O modules, the virtual environment allows quick, easy structural changes by adding or removing a rack and adjusting logic. Once a cabinet or rack has been created by the tool, its configuration is stored in a database, providing configuration management functions. If a newly constructed power plant uses the same hardware version of the PLC and DCS as a previous one, much of the content can be reused to develop a prototype. Of course, several engineering activities are needed for making virtual networks.

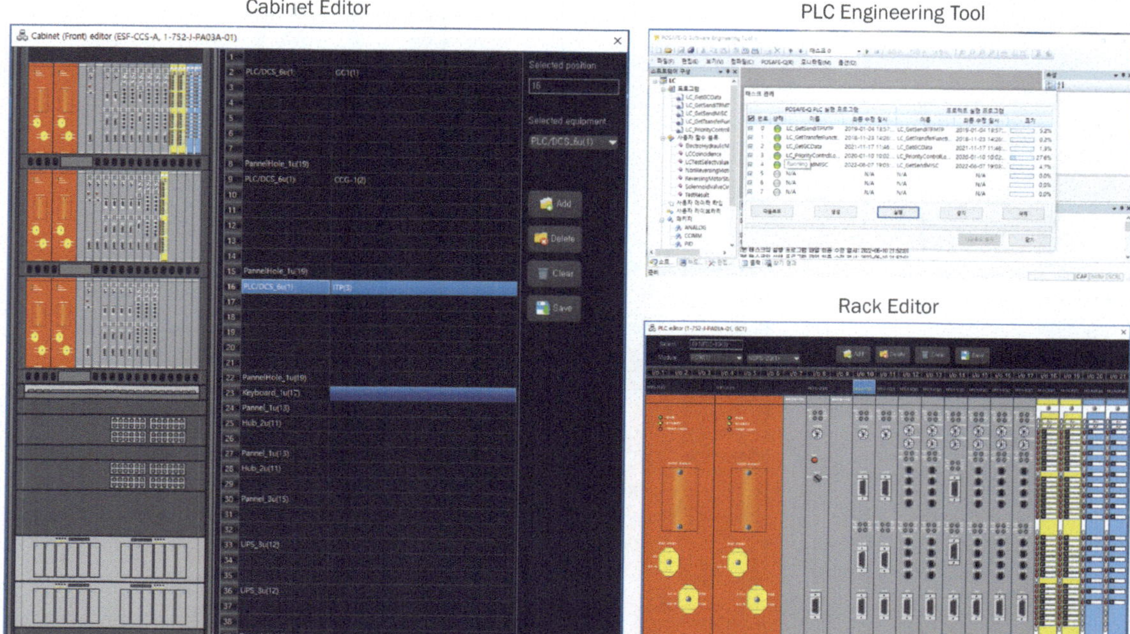

FIG. VII–9. Exclusive engineering tools for the virtualized MMIS (courtesy of S. Lee, Korea Hydro & Nuclear Power Company, Republic of Korea).

Figure VII–10 shows a typical example of the unit test and stability test of platform software and PC based software. As the CPUs and peripheral parts of the processor modules for PLC and DCS are completely emulated, pCOS (Programmable Control Operating System) for PLC and VxWorks for DCS run identically in a virtual machine. Possible faults in pCOS, VxWorks and subprograms can be identified over a very long time period. PC based software, such as gateway programs and data link servers, can be also tested. When the PLC or DCS platform is modified in terms of firmware, the embedded OS itself and its associated binary tasks, function testing and software verification and validation can be performed with the available debugging tools. Debugging multiple PLC racks simultaneously requires substantial human resources in the real world, but in the virtual environment it is straightforward and largely automated. Various debugging conditions can be achieved within the virtualized MMIS.

Rack and cabinet logic I/O test: A rack can be created from a virtual DCS or PLC with its own I/O and control agents. A cabinet comprises several racks and CIMs, remote monitoring units and/or terminal blocks. All the values of each rack or cabinet I/O points can be adjusted within the smart engineering tool, which is an exclusive software tool for the APR1400 MMIS virtual stimulation. Possible faults such as incorrect configuration in I/O points and logic errors can be detected through automated I/O test conditions (see Fig. VII–11).

Function test of subgroup of I&C systems: If there are given inputs and expected outputs for several racks and cabinets of each system, a special graphical user interface can be implemented for the unit test of each system. Differences between the expected outputs and the simulated outputs will be investigated for the unit test. If there are no differences, it shows that a system is well designed and made.

I&C systems integration test: A system integration test can be done using time based malfunctions, remote functions and override functions with plant dynamics models. The expected behaviour of the plant can be identified.

All of the above tests can be automated, helping to find potential faults within a shorter time. Reduced testing time and increased reliability ultimately lead to reducing the cost of the power plant construction in terms of time, human resources and finance.

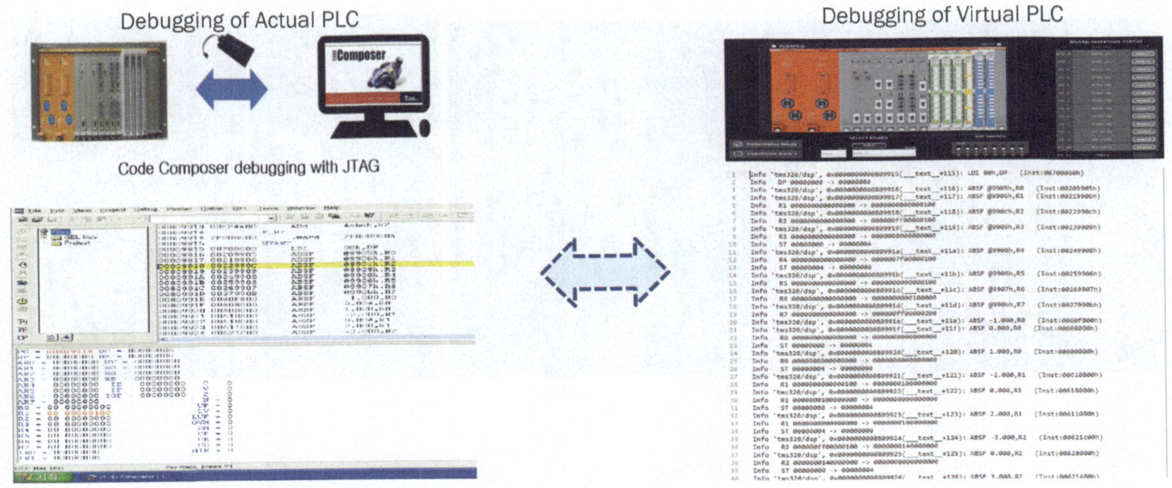

FIG. VII–10. *Software debugging results on a real PLC and a virtual PLC (courtesy of S. Lee, Korea Hydro & Nuclear Power Company, Republic of Korea).*

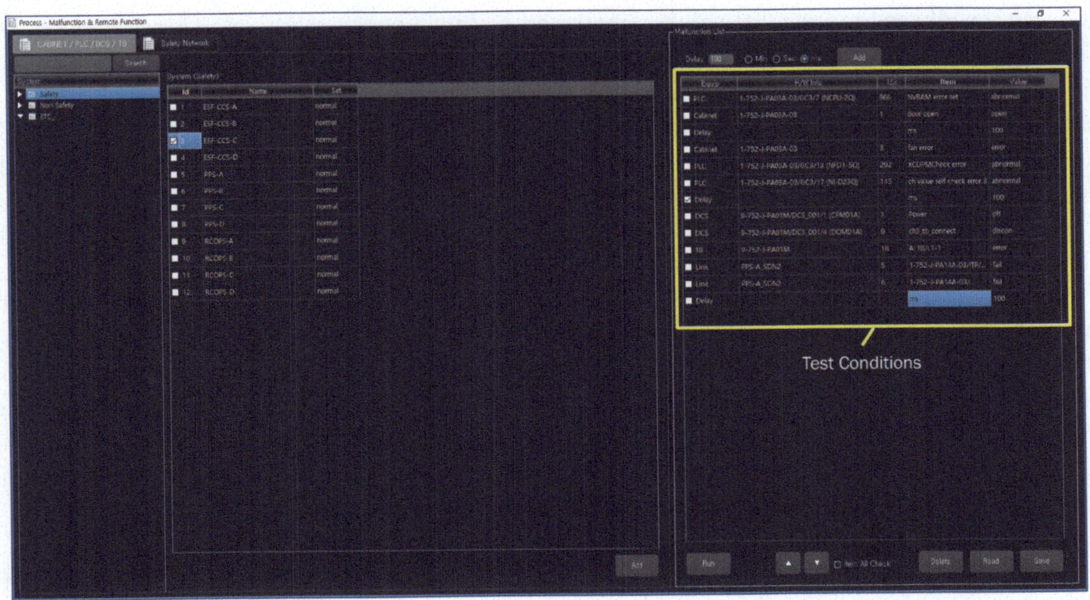

FIG. VII–11. *Screenshot of the malfunction (courtesy of S. Lee, Korea Hydro & Nuclear Power Company, Republic of Korea).*

VII–3.2. Operation and maintenance phase support

I&C equipment and network performance monitoring, as shown in Fig. VII–12: The rack and cabinet's operating status, network load and transmitted data from a source point can be measured at any point, helping maintenance and network engineers better understand the behaviour of the real MMIS. It is usually difficult to measure network performance for a real plant because special analysis hardware and software tools are necessary, but the network packets are captured by the smart engineering tool without those special tools. In addition, the performance indices of all the virtual processor modules are

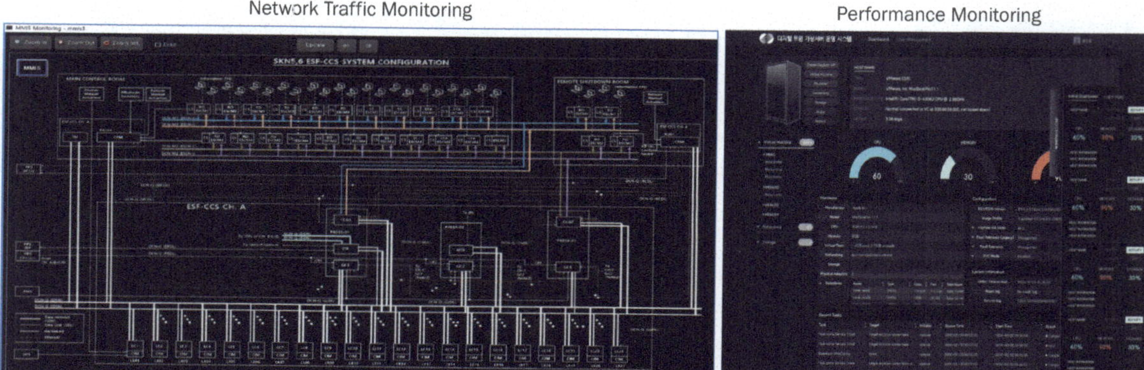

FIG. VII–12. Example of network traffic monitoring (left) and virtual machine performance monitoring (right) (courtesy of S. Lee, Korea Hydro & Nuclear Power Company, Republic of Korea).

monitored, enabling MMIS engineers to understand the overall behaviour and detect abnormal status during the design modification test or tests for other purposes.

MMIS equipment troubleshooting based on its malfunctions: All the components in the virtualized MMIS can be adjusted to intended values, simulating malfunctions or overrides. These intended values are inserted at a specific time according to a test scenario. Various test scenarios can be created by the smart engineering tool. Comparing an abnormal status of the real problem with postulated statuses that are simulated by the various malfunctions in the virtualized MMIS helps identify root causes. If one of the postulated statuses is the same as the real problem, that malfunction may be the cause of the real problem. However, this malfunction editor has been implemented as shown in the right hand part of Fig. VII–11. Malfunctions can be applied at the various levels of the processor module's main subcomponents, all the kinds of I/O modules and communication modules, simulated hardwired cables and so on.

Design modifications of platform software and control logics, including general application software: Before modifications, alternatives will be tested in the virtualized MMIS, from unit testing to integration testing, as well as stability testing. An optimized alternative can be determined based on the preliminary trials. Because the real engineering tools for the virtual PLC and DCS are provided, a user can change the installed control logics on the virtualized environment, or they can replace the old logic files with new ones, and then the necessary tests in terms of function and performance are performed. If the tests are performed twice, before and after the design change, the difference between them can even be determined.

Engineer and operator training facility with intended malfunctions: The virtualized MMIS offers various combinations of displays such as operator displays and MMIS cabinets at the same time. There are many cooperative tasks in the real plant between the MMIS engineer and the main control room operator. One of the typical examples of those tasks is periodic testing of safety I&C systems. During the test, the main control room operator and MMIS engineer have frequent communications and controls as feedback from each other's action. If there is no special training facility, it is very difficult to practise these kinds of tasks for both at the same time. The virtualized MMIS provides all the HMIs of the operator workstation, including a safety console and a large display panel. It also provides the visualization of all the I&C cabinets that are implemented in the virtualization scope. Therefore, the MMIS engineer and main control room operator can manipulate the virtualized MMIS at the same time in order to practise collaborative tasks. Of course, during the manipulation, many malfunctions are injected. This leads to more practical training for the engineer and operator. The virtualized MMIS provides the training modes for the operator or the engineer with a kind of instructor station. For example, when an instructor injects several malfunctions into the virtualized MMIS, trainees use the smart engineering tool to determine where the problems are located.

Real equipment testing: The real DCS and/or PLC cabinets can be integrated with the virtualized MMIS. DCS cabinets are connected to other virtual DCS cabinets via Ethernet cables. The real DCS cabinets have a specific physical tool to send or receive their I/O data. For PLC cabinets, the connection differs slightly, depending on the safety network for the APR1400 MMIS, such as SDN and SDL cables. The SDL is a data link that connects each channel in a safety system, while the SDN is a data network within one channel. This connection is virtualized through an SSGW PC developed for this project.

For an example of a Qualified Indication and Alarm System – Non-safety (QIAS-N), Fig. VII–7 shows the real equipment integration using an SSGW PC and a virtualized QIAS-N. It is noted that data transfer between the two cabinets was successful. This functionality enables validation and verification of the virtualized parts and vice versa. The special I/O tools for the real PLC and DCS should exchange data only with the virtualized MMIS. This ensures that the I/O data can be controlled to perform various tests.

Cybersecurity testing: Once a power plant is constructed or under construction in the real world, a set of physical MMIS cabinets is necessary to perform a cybersecurity test. Therefore, most of the cybersecurity tests have been performed in the later phase of the plant construction. Because of the implementation of the identical network topology, various cybersecurity tests can be performed, such as the penetration test or network hardware malfunction simulations. The virtualization of the real network switch has not yet been developed, so there are some limitations. It is one of the next items planned for development.

VII–3.3. Future applications of the Man–Machine Interface System virtualization simulator

Test bed for MMIS R&D, including artificial intelligence (AI) edge computing and the Internet of Things (IoT): Virtual machines provide resources that can be extended for computational functions. Many state of the art techniques and technologies can be integrated into the virtualized MMIS. Virtual I/O parts and safety networks have been implemented using Ethernet connections, allowing various IoT devices to replace those parts. This enables deployment of virtual controllers through the components and their physical I/O parts based on IoT. It is a kind of hybrid control system for the real plant. Many AI modules or experimental functions can be deployed on the available space as mentioned above, so they can be demonstrated or validated in the virtual environment before real deployment. This approach reduces the time and cost for those kinds of R&D projects. This virtualized MMIS and associated facilities, including space and computational resources with the real plant information, will be used for an R&D test bed for MMIS field stakeholders in the Republic of Korea.

Cloud based virtual control system: Owing to the characteristics of virtual machines, all virtualized MMIS components can be loaded into a private or public cloud system, provided the computational requirements are met.

VII–4. CONCLUSION

The APR1400 MMIS simulator has introduced a new concept known as 'virtual stimulation'. Virtual stimulation means that all PLC and DCS cabinets are virtualized with the same software and simulated hardware. This approach supports almost all engineering activities and tests performed in real I&C systems, as well as new engineering activities and tests in virtual I&C systems — covering both aspects of digital I&C systems. It is expected that this kind of MMIS simulator will serve many versatile purposes, such as plant performance optimization, plant construction cost reduction and improvement of engineering ability for MMIS engineers. Unlike in real I&C systems, where the internal status is difficult to monitor or observe, the internal status of the virtual I&C system can be readily observed. Based on the virtualized MMIS, a smart engineering tool has also been developed to enable various tests to be carried out easily and quickly. It is also possible to perform cybersecurity tests on the virtualized MMIS, because it simulates the real network topology based on the Ethernet connection, SDN/SDL connection and other hardwired connections. Virtualization technology makes it possible to create identical versions of real systems in the software. It is expected that the virtualized MMIS will be used for a wide range of purposes, from fault detection to complex testing and engineering analysis. These activities will help improve the capability of the plant staff and ensure the reliability of I&C facilities in terms of plant performance.

REFERENCES TO ANNEX VII

[VII–1]   ELECTRIC POWER RESEARCH INSTITUTE, Systems Engineering Process – Methods and Tools for Digital Instrumentation and Control Projects, Rep. 3002008018, EPRI, Palo Alto, CA (2016).

[VII–2]   ZHANG, X., et al., Design and verification of reactor power control based on stepped dynamic matrix controller, Sci. Technol. Nucl. Install. **2019** (2019) 4973120, https://doi.org/10.1155/2019/4973120

[VII–3]   LEE, S., KO, W.W., "Basic concepts of APR1400 MMIS digital twin using virtualization technology", Transactions of the Korean Nuclear Society Spring Meeting, Jeju, Republic of Korea, 2020.

GLOSSARY

big data. Data sets that are too large or too complex to be dealt with using traditional data processing methods and tools.

condition monitoring. The observation of an installation and of its systems, components and structures, and analysis of the information collected to detect signs of abnormal behaviours or conditions.

data logging. The recording over time of information collected on an installation and on its systems, components and structures.

diagnostics. A discipline concerned with identifying the cause of a system or a component failing to perform its intended function.

digital twin. A model that serves as a digital counterpart of a physical object or system, as built or to be built and installed on-site, at a level of accuracy suitable for a given engineering activity. Different activities may need different digital twins.

on-line monitoring. An automated method of condition monitoring while the installation, the system or the component is operating.

prognostics. A discipline concerned with predicting when a system or a component will no longer perform its intended function.

surveillance testing. The activity of checking a system or component to determine whether it is operating within acceptable limits.

ABBREVIATIONS

AI	artificial intelligence
ALARA	as low as reasonably achievable
ALFC	advanced load following control
API	application programming interface
AR	augmented reality
BEPU	best estimate plus uncertainty
BIM	building information modelling
BWR	boiling water reactor
CAD	computer aided design
CAM	computer aided manufacturing
CAVE	cave automatic virtual environment
CIM	Common Information Model
CMMS	computerized maintenance management system
CMS	core monitoring system
CPU	central processing unit
CVCS	chemical volume control system
DCS	distributed control system
DIP	data integration platform
EDF	Électricité de France
EPRI	Electric Power Research Institute
FIC	flow induced corrosion
FlexOp	flexible operation
FMEA	failure mode and effects analysis
FMECA	failure mode, effects and criticality analysis
FMSA	failure mode and symptoms analysis
FTC	fault tolerant control
HFE	human factors engineering
HMI	human–machine interface
HSI	human–system interface
HVAC	heating, ventilation and air conditioning
I&C	instrumentation and control
IEC	International Electrotechnical Commission
IFC	Industry Foundation Classes
I/O	input/output
KHNP	Korea Hydro & Nuclear Power
KWU	Kraftwerk Union
LOCA	loss of coolant accident
MMIS	Man–Machine Interface System
MML	Metroscope modelling library
OLM	on-line monitoring
OS	operating system
PC	personal computer
PCI	pellet–cladding interaction
PLC	programmable logic controller
PSA	probabilistic safety assessment
PWR	pressurized water reactor
R&D	research and development

REO	rated electrical output
RMS	root mean square
RPA	robotic process automation
RTOS	real time operating system
RUL	remaining useful life
SDL	safety data link
SDN	safety data network
SIEM	security information and event management
SMR	small modular reactor
SNCF	Société nationale des chemins de fer français
SOAR	security orchestration, automation and response
VR	virtual reality

CONTRIBUTORS TO DRAFTING AND REVIEW

Afonin, E.M.	Rusatom Automated Control Systems, Russian Federation
Bi, D.	Shanghai Nuclear Engineering Research and Design Institute, China
Busquim, R.	International Atomic Energy Agency
Crétinon, L.	Électricité de France, France
Drouet, V.	Metroscope, France
Eiler, J.	International Atomic Energy Agency
El Bouzidi, S.	Canadian Nuclear Laboratories, Canada
Fomi Wamba, F.	Framatome, Germany
Hashemian, H.M.	Analysis and Measurement Services Corporation, United States of America
Lee, S.	Korea Hydro & Nuclear Power Company, Republic of Korea
Li, Y.	China Nuclear Power Engineering Corporation, China
Morokhovskyi, V.	Framatome, Germany
Nekula, P.	CEZ Group, Czech Republic
Ngoy Kubelwa, N.	International Atomic Energy Agency
Nguyen, T.	Private consultant, France
Salnikova, T.	Framatome, Germany
Schneidesch, C.	Tractebel Engineering, Belgium
Schwartz, A.	Metroscope, France
Shumaker, B.	Analysis and Measurement Services Corporation, United States of America
Song, F.	Shanghai Nuclear Engineering Research and Design Institute, China
Wang, Y.	China Nuclear Power Engineering Corporation, China
Weser, F.	Framatome, Germany
Yao, W.	China Nuclear Power Engineering Corporation, China
Zhao, J.	Nuclear Regulatory Commission, United States of America

Consultants Meetings

Vienna, Austria: 22–26 November 2021, 25–29 April 2022, 12–16 December 2022

Technical Meeting

Vienna, Austria: 30 August–2 September 2022

Structure of the IAEA Nuclear Energy Series*

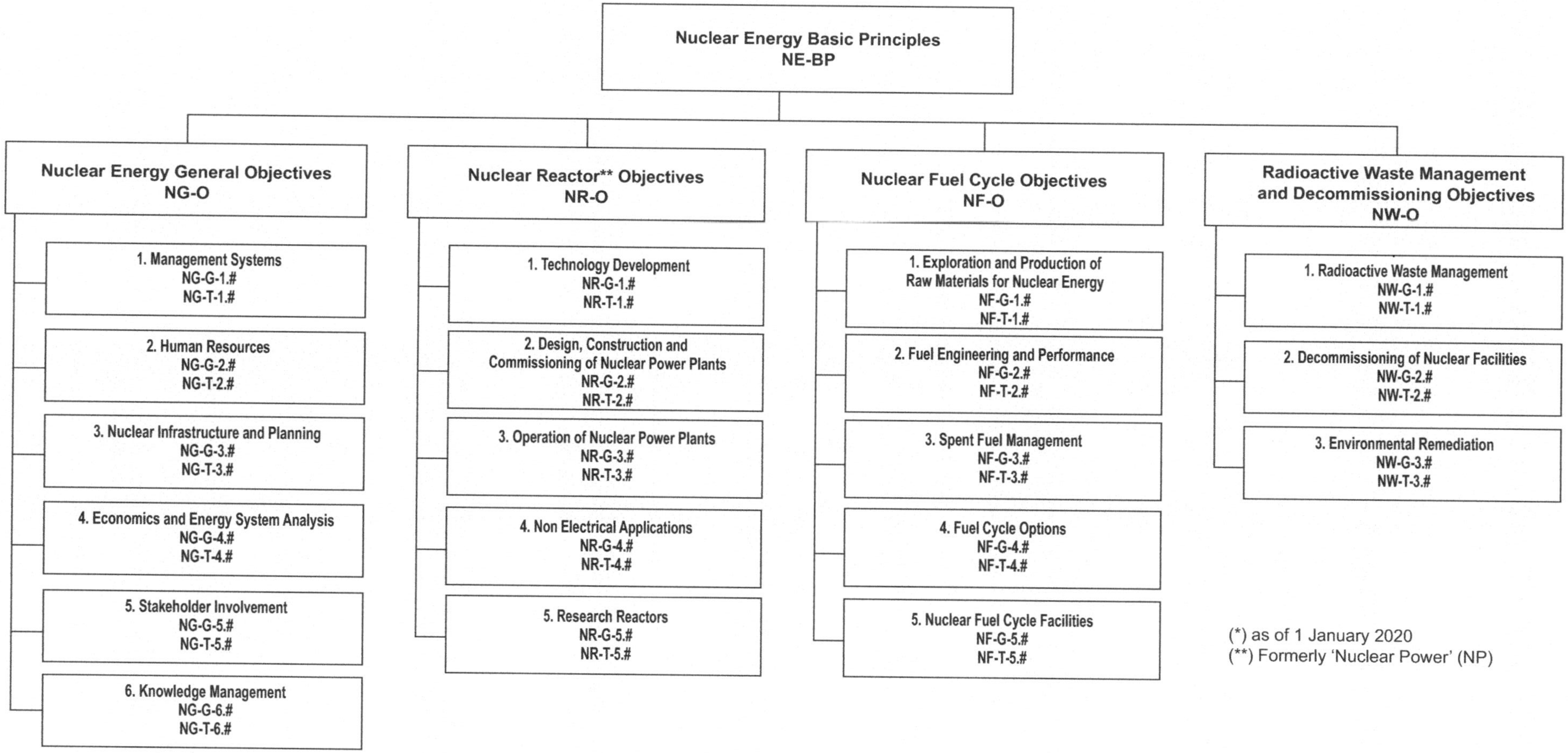

Nuclear Energy Basic Principles
NE-BP

Nuclear Energy General Objectives NG-O

1. Management Systems
NG-G-1.#
NG-T-1.#

2. Human Resources
NG-G-2.#
NG-T-2.#

3. Nuclear Infrastructure and Planning
NG-G-3.#
NG-T-3.#

4. Economics and Energy System Analysis
NG-G-4.#
NG-T-4.#

5. Stakeholder Involvement
NG-G-5.#
NG-T-5.#

6. Knowledge Management
NG-G-6.#
NG-T-6.#

Nuclear Reactor** Objectives NR-O

1. Technology Development
NR-G-1.#
NR-T-1.#

2. Design, Construction and Commissioning of Nuclear Power Plants
NR-G-2.#
NR-T-2.#

3. Operation of Nuclear Power Plants
NR-G-3.#
NR-T-3.#

4. Non Electrical Applications
NR-G-4.#
NR-T-4.#

5. Research Reactors
NR-G-5.#
NR-T-5.#

Nuclear Fuel Cycle Objectives NF-O

1. Exploration and Production of Raw Materials for Nuclear Energy
NF-G-1.#
NF-T-1.#

2. Fuel Engineering and Performance
NF-G-2.#
NF-T-2.#

3. Spent Fuel Management
NF-G-3.#
NF-T-3.#

4. Fuel Cycle Options
NF-G-4.#
NF-T-4.#

5. Nuclear Fuel Cycle Facilities
NF-G-5.#
NF-T-5.#

Radioactive Waste Management and Decommissioning Objectives NW-O

1. Radioactive Waste Management
NW-G-1.#
NW-T-1.#

2. Decommissioning of Nuclear Facilities
NW-G-2.#
NW-T-2.#

3. Environmental Remediation
NW-G-3.#
NW-T-3.#

(*) as of 1 January 2020
(**) Formerly 'Nuclear Power' (NP)

Key
BP: Basic Principles
O: Objectives
G: Guides and Methodologies
T: Technical Reports
Nos 1–6: Topic designations
#: Guide or Report number

Examples
NG-G-3.1: Nuclear Energy General (**NG**), Guides and Methodologies (**G**), Nuclear Infrastructure and Planning (topic **3**), **#1**
NR-T-5.4: Nuclear Reactors (**NR**), Technical Report (**T**), Research Reactors (topic **5**), **#4**
NF-T-3.6: Nuclear Fuel (**NF**), Technical Report (**T**), Spent Fuel Management (topic **3**), **#6**
NW-G-1.1: Radioactive Waste Management and Decommissioning (**NW**), Guides and Methodologies (**G**), Radioactive Waste Management (topic **1**) **#1**

Feedback on IAEA publications may be given via the on-line form available at:
www.iaea.org/publications/feedback
The form may also be used to report safety issues or submit environmental queries concerning
IAEA publications.

Alternatively, contact IAEA Publishing directly:

Publishing Section, Dissemination Unit
International Atomic Energy Agency
Vienna International Centre, PO Box 100, 1400 Vienna, Austria
Telephone: +43 1 2600 22529 or 22530
Email: sales.publications@iaea.org
www.iaea.org/publications

ORDERING LOCALLY

To purchase priced IAEA publications, please contact either your preferred local supplier
or the IAEA's lead distributor:

Mare Nostrum Group
39 East Parade
Harrogate
North Yorkshire
HG1 5LQ
United Kingdom
Email: enquiries@mare-nostrum.co.uk
www.mngbookshop.co.uk

Trade Orders and Enquiries:
Telephone: +44 1243 843 291
Email: trade@wiley.com

Individual Orders and Enquiries:
Email: mng.csd@wiley.com

Priced and unpriced IAEA publications may also be ordered directly from the IAEA by contacting
IAEA Publishing. The recipient is responsible for shipping costs and any customs and duties.

Printed and bound by CPI Group (UK) Ltd, Croydon, CR0 4YY

06/07/2026

02160600-0013